cat care ess

D0756010

hamlyn | **all colour petcare**

cat care essentials

Francesca Riccomini

hamlyn

An Hachette UK Company
www.hachette.co.uk

First published in Great Britain in 2010 by
Hamlyn, a division of Octopus Publishing Group Ltd
Endeavour House
189 Shaftesbury Avenue
London
WC2H 8JY
www.octopusbooks.co.uk
www.octopusbooksusa.com

Copyright © Octopus Publishing Group Ltd 2010

Distributed in the U.S. and Canada by
Octopus Books USA:
c/o Hachette Book Group
237 Park Avenue
New York, NY 10017

ISBN 978-0-600-62056-3

A CIP catalogue record for this book is available from the British Library

Printed and bound in China

10 9 8 7 6 5 4 3 2 1

Note Unless the information is specific to males or females, throughout this book kittens and cats are referred to as 'he' and the information and advice are applicable to both sexes.

The advice in this book is provided as general information only. It is not necessarily specific to any individual case and is not a substitute for the guidance and advice provided by a licensed veterinary practitioner consulted in any particular situation. Octopus Publishing Group accepts no liability or responsibility for any consequences resulting from the use of or reliance upon the information contained herein. No cats were harmed in the making of this book.

Contents

Introduction

Confident, cute, curious, calm, complacent – whatever an individual cat's personality or mood of the moment, the chances are that he's stolen somebody's heart!

More people than ever now recognize the benefits that living with cats can bring. These elegant and appealing creatures invariably entertain, intrigue, comfort, and even inspire, the folks upon whom they deign to bestow their attention. No wonder few cat lovers feel their home is complete without a familiar feline presence somewhere about the place.

The flip side is that our responsibility for providing everything they need to keep our cats happy, healthy and safe can be daunting. Whether we are experienced cat people or novice owners, increased understanding and up-to-date knowledge always help. The aim of this quick reference guide is to make the task not only easier but also more interesting and enjoyable than it might otherwise be.

Cat Care Essentials will give you an insight into feline behaviour, physical make-up, emotional needs and health-care tips, so that you will be well on your way to providing a cat-friendly home, with all that your cat needs for comfort, entertainment and a peaceful existence. Although two cats are never the same, their common requirements are covered in detail and answers given to everyday questions that may puzzle you from time to time. You can look forward to sharing your life with a contented feline companion and forging a great relationship.

Cat facts

Your cat has changed little since his ancestors learned to survive the harsh savannah environment, and it is his inherited characteristics that underpin his day-to-day activities and his physical and emotional needs. This chapter gives you a brief insight into what makes your cat the unique pet that you welcome into your home.

Family matters

A little understanding of his ancestral inheritance will help you appreciate and understand all the little quirks and foibles that make cats so appealing, so you can live contentedly side by side despite your different species characteristics.

Domestication – your cat is descended from the African wild cat, *Felis silvestris*, although exactly when domestication of this shy, solitary species took place is still debated. However, it is generally recognized that once the Egyptians stored their cultivated grain, the value of their feline neighbours as rodent controllers quickly became apparent. From the cats' perspective, a reliable supply of food and shelter had considerable advantages, so the relationship began and has prospered ever since.

WILD CAT FACTS

Food, shelter and survival are any cat's top priority. The table below illustrates how survival in the wild translates into a happy home life for your domestic feline companion:

Wild cat	Your cat
Solitary living – cats seek each other only for mating, then the mother raises her kittens alone	Probably happy to live independently without a feline companion
Highly developed for nocturnal hunting – most active at dawn and dusk (crepuscular) when prey is out and about	Spends considerable time asleep during the day but is still more awake in the early morning and evening
Stalking and hunting prey	Playing pounce and chase games
Free-ranging and territorial – an individual's existence depends upon the territory he can lay claim to and what it contains	Scent-marking in the garden and in the house to establish his territory
Active defence is risky – communication between cats is used to increase not reduce the distance between them	Avoids confrontation as much as possible and needs safe refuges, often high up, to escape to and observe from

Wild at heart – surprisingly though, despite coming a long way together, we humans have had relatively little influence upon our feline companions, especially compared to our other popular pet, the dog. With the exception of pedigree breeds, even pet cats have simply gone about the business of choosing mates and breeding themselves. As a result, their ancestral survival instincts still govern their behaviour. By recognizing and understanding these basic needs, you can avoid potential difficulties or stress for your feline pet, and gain maximum enjoyment from living with him (see box).

Feline design

Some differences and biological adaptations explain feline grace and athleticism but, apart from size, the cat that lives with you differs surprisingly little in build and behaviour from the big cats in the wild.

FUR FACTS

A cat's fur has a number of functions: insulation; camouflage; and protection of the skin. The coat and skin are made up of:

Guard hairs form the longer topcoat

Awn hairs are bristly and comprise the middle coat

Down hairs are shorter, curly and soft and comprise the undercoat

Whiskers, or vibrissae, are specialized hairs around the face and on the forelegs that are used as sensors

Specialized skin glands on the chin, around the mouth, base of the ears, anus, and between the toes, produce scent secretions that have an important role in feline communication

Body shape – varies in domestic cats depending upon their origin. Those from hotter climes have a more slender build than cats hailing from colder regions, where a stockier build is more advantageous.

Face shape – is also different, with a narrower skull and pointed chin

KEY QUESTION

Why does my cat have such long whiskers?

Cats use their whiskers to gather information about the environment, for detecting air currents and vibrations and when navigating between objects, even in gloomy conditions. Having long whiskers is particularly useful when judging distances in confined spaces, the longer they are the less likely your cat is to bump into things. The position of your cat's whiskers will also help you to work out how he is feeling (see page 28).

Bones in the feline body

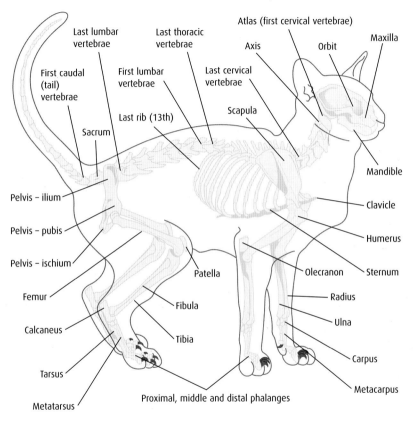

Last lumbar vertebrae

Last thoracic vertebrae

Atlas (first cervical vertebrae)

Axis

Orbit

Maxilla

First caudal (tail) vertebrae

First lumbar vertebrae

Last cervical vertebrae

Last rib (13th)

Scapula

Sacrum

Mandible

Pelvis – ilium

Clavicle

Pelvis – pubis

Humerus

Pelvis – ischium

Patella

Olecranon

Sternum

Femur

Fibula

Radius

Calcaneus

Tibia

Ulna

Tarsus

Carpus

Metatarsus

Proximal, middle and distal phalanges

Metacarpus

distinguishing oriental types from the more rounded profile of their northern cousins.

Paws and claws – all cats have retractable claws so that a feline paw print can be distinguished from a dog's by the latter's claw marks. Sweat glands are only found in the skin of a cat's paw pads, which is why they leave wet paw marks behind when they are anxious.

Movement and balance

Whilst they are not indestructible, cats do have an enviable ability to balance in the most precarious circumstances, and to twist and right themselves to land safely if they fall. Nervous control, the workings of the inner ear and the useful counter-balance provided by the tail are responsible for this feline skill.

The central nervous system – is made up of the brain and spinal cord. The brain processes information relayed to it from the main sense organs – the eyes, ears and nose – and the peripheral nervous system. In turn it influences all the body's systems and the animal's behaviour via these peripheral nerves and hormones, which are secreted by glands, such as the thyroid and adrenals, and organs, including the female's ovaries and the male's testicles.

KEY FACT

'High rise syndrome' is a phenomenon where cats can sometimes survive falls relatively unscathed so long as they have time to right themselves, and don't tense up so rigidly that their legs absorb the full impact when they hit the ground. However, sadly, many cats are badly injured in falls from windows and balconies.

The autonomic nervous system – is an involuntary system that controls unconscious actions, such as blood circulation, breathing and digestion. Consisting of two parts, the sympathetic system, which is responsible for arousal, and the parasympathetic, involved in relaxation and return to normal (homeostasis), it is important in the 'flight or fight' reactions that help keep cats out of trouble whenever possible.

The righting reflex – the vestibular apparatus in the inner ear consists of fluid-filled canals lined with sensitive hairs, which detect fluid movements when the head changes position. This combined with visual information is relayed rapidly to the area of the brain involved with posture, the cerebellum. As a result, when falling a cat can twist his head into a normal position relative to the ground and his body will spontaneously move to correct itself.

Circulatory system

Your cat's activity pattern – long periods of relaxation intermingled with bursts of intense action – would be impossible without effective circulatory and respiratory systems. With the need to be instantly alert and on the move, your cat has developed breathing and heart systems that are primed for flight.

Blood – carries oxygen and nutrients around the body to all the organs, including the lungs, brain, liver and kidneys, as well as supplying the tissues with various life-sustaining functions. Then it helps in the removal of waste products and essential detoxification processes.

The vascular system – is a complicated network of thicker-walled, elastic arteries and thinner veins that delivers blood to the target organs and tissues and returns it to the heart, which acts as a pump driving the whole system. The left side of the heart pumps oxygenated blood around the body, delivering oxygen and nutrients. Then, after returning via the venous system to the right cardiac side, the deoxygenated blood is pumped with the next heartbeat to the lungs, where carbon dioxide is exchanged for more oxygen taken in with each breath.

Pulse rate – is created by the pump action of the heart. The resting feline heart rate of around 120 beats per minute is higher than ours and it can almost double with a speedy burst of activity.

Respiratory system – like humans, cats have upper and lower parts to their respiratory system. Scents are detected when inhaled air passes through the nostrils, then the sinuses where it is warmed and fine hairs filter out

BLOOD GROUPS

Cats have A, B and AB blood groups, although the latter is rare. As blood group is genetically determined, the incidence of the different types can vary both geographically and with breed.

The circulatory system

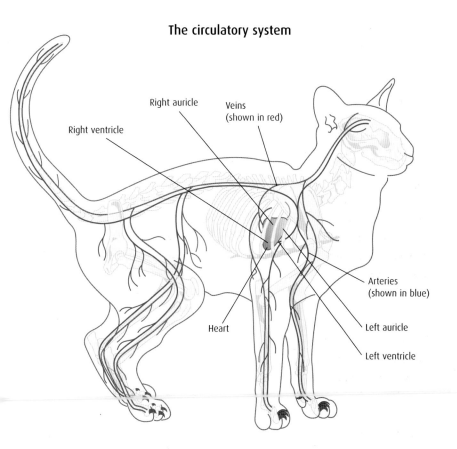

Right auricle

Veins
(shown in red)

Right ventricle

Arteries
(shown in blue)

Left auricle

Heart

Left ventricle

particles. Its subsequent journey to the lungs via the trachea, bronchi and tiny bronchioles allows blood vessels to replenish the body's oxygen supply and eliminate waste. At around 40 breaths per minute the feline resting respiratory rate is again higher than the human rate. A cat's respiratory rate increases with exertion but will quickly return to normal in a fit and healthy cat. Cats rarely pant unless they are ill or very agitated, because they sweat through the pads on their paws.

The digestive system

Cats that fend for themselves are exceedingly skilful predators but even well-catered-for pets are amazingly well adapted to their role as carnivores. It is important to provide a good diet with the correct nutrients and plenty of fresh water to keep your cat's digestive system in top working order.

Carnivores – cats require a high protein diet derived from meat and have become so specialized that they need a constant supply of the right proteins (see page 128) or they can develop health problems associated with continued liver activity.

Teeth – as he grows to adulthood, a kitten's 26 milk (temporary) teeth are replaced by 30 permanent teeth:

- 4 canines or fangs – catching and dispatching prey
- 12 incisors (tiny front teeth) – grip food and are used in grooming
- 10 premolars (3 each side at the top and 2 in the lower jaw) and 4 molars – shearing, tearing and chewing, although cats tend to quickly swallow their food

Tongue – the surface of your cat's tongue is rough because of the papillae that are located in a band down the middle. These barbs point backwards and help to pull off and grip pieces of meat when he's eating. During grooming they also act to comb his fur. When your cat drinks, he scoops up the liquid by curling back the tip of his tongue to form a hollow.

The skull and teeth

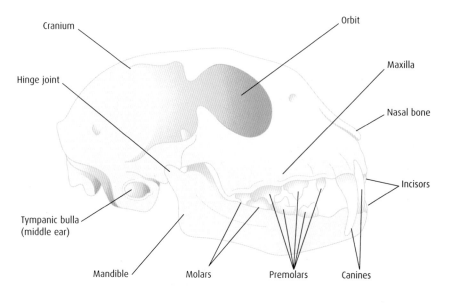

Cranium

Orbit

Hinge joint

Maxilla

Nasal bone

Incisors

Tympanic bulla
(middle ear)

Mandible

Molars

Premolars

Canines

Saliva – starts the process of digestion in the mouth and mixes with food to ease its passage down the oesophagus (gullet) into the stomach.

Digestion – acid and enzymes in the stomach continue the process of breaking down meat into its constituent components before it is propelled by muscular action into the intestines. Digestion continues in the small intestine (duodenum), which is

encountered first, but ceases when the larger intestine (colon) is reached. Water is now reabsorbed before the resultant faeces enter the rectum prior to elimination.

Nutrition – once they pass through the intestinal wall a well-developed blood supply carries the nutrient end products of digestion to the liver, where they are further broken down into amino and fatty acids that fuel the body.

Feline senses

It is impossible for us with our human perspective to understand how differently cats view the world. Some of their skills are superior to ours, which can make living in our homes with noisy electrical gadgets and strong, artificial fragrances, somewhat challenging.

Smell – the membrane containing odour-detecting cells, lining the cat's nose, is twice as extensive as ours. Another advantage they have is a specialized pouch-like structure in the roof of the mouth, the vomeronasal or Jacobson's organ, which enables cats to taste scents and gain extra information (see Funny faces, left).

Sight – like us, cats have binocular vision but a wider field of view. Their pupils also have a greater ability to contract and dilate in response to light. A layer of specialized reflective cells lining the back of the eye (the tapetum lucidum) allows the retina to absorb extra light, thus aiding night vision, and causing cats' eyes to flash in the dark. Despite this, they cannot see in complete darkness.

Hearing – the size, shape and extraordinary flexibility of a cat's ears, which can move independently of each other to increase effectiveness, allows detection of sounds

FUNNY FACES

You may notice your cat pulling a funny face, with lips curled back and mouth partly open in a mesmerized grimace. This is known as the Flehmen response and occurs when your cat is 'tasting' the air, trying to gather as much information as possible from a scent, especially if another cat has entered his territory and left odour signals on plants or furniture.

The brain and its functions

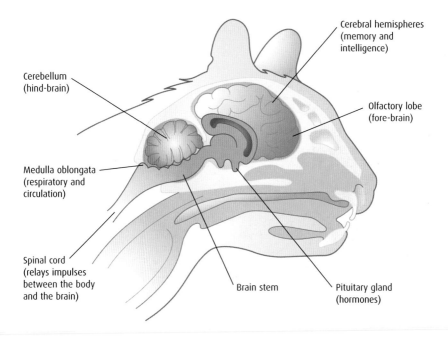

Cerebral hemispheres (memory and intelligence)

Cerebellum (hind-brain)

Olfactory lobe (fore-brain)

Medulla oblongata (respiratory and circulation)

Spinal cord (relays impulses between the body and the brain)

Brain stem

Pituitary gland (hormones)

over a wide area. The frequency range over which cats hear is much greater than ours, allowing them to detect and distinguish rodent movements over 9 m (30 ft) away.

Taste – this is difficult to assess in cats, and as meat eaters, they have little ability to detect sweetness but can taste salt, bitter and sour flavours using taste buds on the tip, sides and back of the tongue.

Touch – their vibration- and temperature-sensitive whiskers help cats detect prey and negotiate confined spaces but they are also a good indicator of emotional state (see pages 28–29).

Your cat's behaviour

Cats are naturally sensitive creatures, governed by strong instincts that need special handling in order to live stress-free lives with their chosen human companions. With a little understanding and consideration of their needs, you can provide a home where your cat can indulge in a carefree existence.

The feline personality

Every cat is unique but, as many cat owners will tell you, the one personality trait that most of them share is independence. They choose to live with you, not the other way around! Your cat may be aloof, friendly, cuddly, unpredictable, biddable and distant – all the things that make them such an appealing and rewarding pet.

Aloof – your cat's ancestors were originally solitary. They gradually adapted to live alongside people and be cared for in our homes, but most cats still need to maintain their independence. Ensuring your cat has his own space is the best way to create harmony.

Independent – groups of cats in the wild are usually made up of related females, who communally rear and defend kittens as well as repelling territorial invaders. Many cats will prefer their own company, but if you would like more than one pet, consider taking a mother and her kitten, or a brother and sister, for a happy family.

Tolerant – as our cats' ancestors adapted to live alongside people, they also became more tolerant of each other and other animals, including pets. Understanding the dynamics of social interactions and shared living space, will help your cat live peacefully and sociably in your household (see page 82).

Loving – despite their independence, many cats do enjoy gentle handling, cuddles and stroking, rewarding you with a purr of contentment and accepting you as part of their world.

Playful – with their hunting instincts still very much guiding their activities, domestic cats need lots of suitable toys to stalk, chase and catch (see pages 138–139).

Inherited – your cat's parents equally pass on their personal and family traits to their offspring. Boldness, timidity, reactivity and sociability may, to some extent, be inherited.

Experience – some of your cat's behaviour will stem from his early experiences as a kitten. Those kittens used to human handling from a very early age are likely to be more sociable and relaxed in the family home than kittens raised only by one person, left to fend for themselves or with bad experiences of humans.

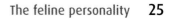

Communication

The wild cats of the African savannah developed communication systems that were designed to maintain, or increase, distance between individuals. Your cat will use these methods to communicate with you and others in their home territory.

Silent treatment – it seems that cat voices are individually unique like ours, but obviously, unless searching for mates, their ancestors wanted to make as little noise as possible for fear of drawing attention to themselves. As a result, many pet cats are relatively silent.

Vocal interactions – are often confined to kittenhood, when mothers announce their return to the nest by gently chirruping to the kittens, or offspring call out in distress if separated from the queen and their littermates. Growling and hissing are used in certain circumstances and their meaning is often clear.

Chatty cats – some pets, however, are remarkably vocal, especially if we respond positively, chatting back in silly voices as many of us do, when feeding, petting or playing with our cats. Learned behaviour of this sort can be quite charming and all we

PURRING

This is the form of vocalization that many people most associate with cats. The exact mechanism by which the purr is generated is still to be identified but from earliest kittenhood the cat's purr gains attention. Together with their tiny front paws paddling their mother's belly, it is the means by which her offspring induce the milk let-down reflex enabling them to suckle. In adulthood, many cats purr when relaxed and contented and, when in response to close contact with us, it helps to bond pets and their owners. Sadly, cats will also purr in conditions of extreme distress, for instance if they are very ill, stressed or injured.

are really doing is building on the normal social greetings that well-bonded, usually related cats reserve for each other when they meet up after an absence.

Communication repertoire – with silence being preferred, cats have a range of other communication methods to call upon, using visual signals and scent markings to show others how they feel. See pages 28–33 for more on these feline traits. However, cats are not so skilled at defusing tension with other felines when their distancing tactics fail.

Body language

A species that wanted to keep others at arm's length needed to communicate at a distance. Consequently, cats use their body's silhouette to convey how they are feeling. There are other clues that also indicate your cat's mood.

Facial expression – hissing, spitting, growling and snarling are vocal signals that affect a cat's facial muscles and are easy to read.

Whisker position – these specially stiffened, sensitive hairs are mobile so they can be folded away and protected if danger threatens. With whiskers back and ears flat, your cat is feeling highly defensive.

Pupil constriction or dilation – unless it is simply a response to very bright (constriction) or low intensity (dilation) light, the way a cat's pupils look tells us a lot. Anxious, excited or fearful cats have 'big eyes'; narrowed, constricted pupils are associated with potential aggression.

Tail position and movement – a straight up, kinked-tip tail is a greeting reserved for familiar individuals, whereas a flicking tip or swishing tail indicates arousal or irritation.

KEY TIP

Always bear in mind that it is the whole package that counts and taking one signal in isolation can be very misleading.

READING YOUR CAT'S BODY LANGUAGE

Body position	Your cat's mood
Outstretched, relaxed, lying on his side, droopy tail	A contented cat
Upright, relaxed stance, ears up, moderate or brisk walking pace	Relaxed, confident cat
Controlled, vigilant pose, slowly walking forward in a slightly crouched manner	Hunting or approaching a potential threat
Tense muscles, low crouch, ears back, tail tucked up and under	Anxious cat
'Witch's or Halloween' cat with arched back, fur raised (piloerection), bottle brush tail, all fluffed out and curled into a downward U or tucked away to avoid injury	Fearful cat that is likely to attack defensively if challenged, cornered or picked up
Tense, rolled onto his back to expose his belly (very different from the relaxed and happy cat that's rolling around in the sunshine or a nice warm bed)	Aroused cat that is preparing to attack with all five of his defensive weapons – four sets of claws and his fearsome teeth; don't be fooled into thinking he's being submissive
Tall, tense, raised rump, tail extended, forward-pointing ears, stiff forwardly extended whiskers, constricted pupils	Confident, intimidating cat – take care!

Scent messages

With their highly developed ability to detect odours (see page 20) it makes perfect sense for cats to communicate with each other using scent (olfactory) messages. These remain wherever they have been deposited to pass on important information, long after the 'signaller' has passed by and gone on his way.

Common ground – by pushing certain facial and body areas that contain scent glands against people, other pets to which they are well bonded and inanimate items, cats deposit informative odours. These make everyone and everything smell familiar, non-threatening and part of their social group or territorial domain.

Individual scents – specialized skin glands on different parts of your cat's body produce individual scents, which convey different information. For more on urine spraying, see pages 32–33.

Avoidance tactics – leaving scent markings allows cats to transmit information without actually having any contact with each other if they prefer to avoid meeting.

Staking a claim – leaving a scent helps increase a resident cat's confidence if what he 'owns' is securely labelled as his, which is why cats often scent mark more vigorously if they feel unsettled, stressed or threatened.

Understanding – catering for your cat's natural behaviour is essential for his emotional well-being. It will also help you to protect your furniture and fittings.

SCENT MARKING

There are several ways that cats use scent to mark their territory, objects and other individuals:

Bunting with their faces helps cats transfer scents from skin between the eye and ear, the corners of their mouth and chin

Flank rubbing is a means of cats from the same social group creating a communal odour by exchanging their individual scents. They also tend to rub around the legs of people with whom they are familiar and feel quite comfortable

Tail base sniffing might seem fairly unattractive from the human perspective but it is an important aspect of bonding and social recognition for cats. From the scents produced in this area cats may well be able to tell a lot about who's related to whom, the other cat's health status and emotional state

Footpads contain scent glands that leave odours on objects when cats strop, or scratch, with their front claws. The characteristic noise the activity makes is also used to transmit ownership to any cats in the vicinity

Marking behaviour

Sexually active cats advertise their availability and willingness to mate not only vocally – the classic caterwauling that keeps everyone awake at night – but also by spraying urine in strategic locations where other cats will find the information.

Urine markers – both male and female neutered cats also use urine spraying at the periphery of their home range and along trails within it to lay claim to territory, and in their hunting area to convey information about their presence, health status and identity.

Spray posture – there is a specific posture associated with this activity and it is not uncommon for owners to see their pet backed up against a vertical surface, such as a wall, tree or fence post, with a concentrated expression on his face. With his tail held straight and high and hind feet paddling, a cat squirts a few drops of pungent, sticky urine onto the surface. Other cats may later be spotted sniffing carefully and perhaps showing the typical Flehmen 'grin' as they use the vomeronasal organ to gain more olfactory information (see page 20). Cats may also adopt this posture when aroused or excited but they don't spray urine or scent mark at such times.

Middening – although a less common marking behaviour, middening is the habit some cats have of depositing faeces as a

KEY QUESTION

My cat has left faeces in the middle of my bed. What is causing this behaviour?

This may be middening (see above), however, confusion can arise between this and the situation when cats relieve themselves in unacceptable places because they will not go to find the litter tray or outside to the toilet. Quiet areas of the house where they feel safe are then used and if your cat is very timid or stressed, he may feel happier where your scent is strong, and when your bedroom is unoccupied it becomes an ideal latrine area. Give him a litter tray somewhere private that's easy for him to get to and provide lots of safe places for him to hide on the way to his toilet area (see also pages 74–75).

territorial marker. This behavioural trait is very different from the usual fastidious and discreet feline toileting habits, where excreta are carefully covered in a specially prepared hole. Here the faeces is left uncovered in exposed locations such as the roofs of sheds and garages, atop walls and in the middle of lawns or pathways. Cats that behave in this way are likely to be responding to pressure from a high feline population.

Early experiences

Once we realize that our cat's behaviour is a product of the combined effects of inherited behavioural traits, his personal characteristics and experience, it is clear that those responsible for his early weeks bear an awesome responsibility.

CHECKLIST OF EARLY EXPERIENCES

Gentle handling by people – both sexes, all ages, differing appearances

Appropriate exposure to an interesting environment, which contains everything that is normally found in our homes, such as the television and vacuum cleaner

Controlled and non-stressful encounters with other pets (see pages 84–87)

A range of toys, safe and interesting exercise facilities, suitable scratching and toilet facilities (see pages 60–63)

Daily care inspections – checking ears, eyes, claws and bottoms

---

Socialization – kittens are learning before they are even born. However, during the very early weeks of life, physical and neurological development is marching along at a remarkable pace. At this stage they take in and log for future reference an astonishing amount of information that they learn about themselves and those who surround them, two- and four-legged.

Habituation – refers to the way any animal becomes accustomed to potentially frightening things, such as electrical appliances and family activities, that happen all around, if they are presented in a low-key and non-intimidating manner.

Ready to learn – for our cats the socialization and habituation period is two to seven weeks of age. This does not mean that they are unable to learn later in life if they encounter something novel or potentially scary, but their ability to learn is never likely

to be as sensitive or rapid again. Therefore it is important to establish correct handling in a suitable environment from the start.

Life experiences – making sure that your kitten experiences a wide range of situations, from meeting strangers and children to preparing for trips to the vet, from the minute he is born can make an enormous difference to the quality of life cats later enjoy and to their suitability as pets.

KEY FACT

Studies have shown that it doesn't take a lot of gentle handling to produce a well-balanced pet. A minimum of four people is needed, but they must be male and female of different ages, including children. If they handle the kitten daily for at least 40 minutes in total, he is likely to be happy socializing with anyone in future.

Hunting instincts

Watching your cat play with a toy mouse, a length of string or a feathery 'fishing rod' is highly entertaining but your pet is actually fulfilling a basic instinct to hunt. Understanding this behaviour can help to avoid conflict in the home.

Early training – like all skills, predatory behaviour must be refined and honed through observation, learning and experiment. Mother cats will bring back their kills for kittens to play with and practise their survival skills. Even without any apparent tuition, some cats are extraordinarily dedicated and efficient hunters.

Thrill of the chase – you may wonder why your well-fed cat still feels the need to go out and catch mice or birds. His natural instincts are very strong and he will be unable to resist any opportunity that presents itself. However, pet cats are often proficient only to a point and many can't master the killing bite; they may also appear to be playing with their prey, but by making it move, it is easier for them to locate.

Trophies – many owners have woken to find small 'gifts' of prey presented to them. No one is really sure why cats do this but the

KEY QUESTION

Why does my cat 'chatter' when watching birds?

When cats can see prey moving but they can't get to it, they often become quite frustrated and this sort of chattering provides an outlet for all that pent-up energy.

most likely explanation is that they return to their core area – their home – where they feel safest to eat, or where they can store it for later. Never punish your cat for this gift; just accept that this is his nature.

Practise – youngsters and indoor cats are particularly at risk of becoming bored and frustrated if they do not have appropriate outlets for their hunting instincts. Playing with your cat can help by enabling him to practise the various aspects of the hunting sequence:

- Watch
- Stalk
- Chase
- Pounce
- Catch

However, because few of our pets have to feed themselves, they fail to string them together to end speedily in a neat kill. Nevertheless, adequately catering for your cat's natural predatory instincts is essential.

Hunting instincts 37

Territorial issues

It can seem perplexing that, despite being neutered and having everything they need provided for them, many cats establish and defend territory with immense dedication. Others never actively engage in hostilities but several times each day check out small changes, respond to territorial incursions by neighbouring cats and renew their 'ownership' markers.

Following instinct – it is impossible to predict how territorial an individual will be, although neutering does help, as de-sexed cats tend to have smaller territories than their intact counterparts. However, it is important to understand that territorial battles are not waged out of malicious intent but are part of your cat's ancestral make-up.

TERRITORIAL AREAS

A cat's world is divided into:

Core area – where he eats, sleeps, plays and rests

Home range – a buffer between this safe core and other cats outside

Hunting range – often shared with neighbouring pets

Evidently, how much space each cat can call his own is generally determined by factors beyond his control, such as:

- How well established and/or territorial his feline neighbours are
- The number of cats living locally
- How deep-seated his own need to lay claim to space and defend it is

By making sure we provide suitable marking facilities, such as scratching posts, in adequate numbers we can help our cats feel more secure. This reduces the likelihood of our furniture and fittings being damaged but only if they are located where our cats need them, for example territorial boundaries and entrances/exits to the home, where cats feel most vulnerable and need to 'advertise ownership'.

Home ground – knowledge of natural feline behaviour and how a cat's territory is divided up can also help us to deal sensitively with both physical and social changes in our homes and immediate vicinity. Otherwise these can profoundly disturb our cats, often causing an upsurge of bunting, rubbing, scratching and even indoor urinary spraying, which is normally confined to the outdoors. Cats should never be punished for such behaviour.

Creating a cat-friendly home

Cats have a wide range of daily activities that they carry out to feel safe and secure in their home environment. It is up to us, as their owners, to ensure that we recognize their needs and provide all the facilities that they require for their normal behaviour.

Peace and quiet – cats will naturally seek out quiet, secluded places to sleep or rest, whether inside or outside. Many people mistakenly think that because their cats are allowed outdoors they do not need dedicated facilities at home. It is important to provide dark sanctuaries and easily accessible, high places, for both indoor-only and outdoors cats. Evidently, if cats are indoors all the time they need more plentiful resources devoted to them.

Feeding – a dedicated area where your cat can find their food and a fresh supply of water is essential.

Toileting facilities – it is also important to ensure that our cats' latrine facilities are appropriate, not only in design and cleanliness but also location. Cats cannot easily compromise so you must get things right. In normal circumstances they eliminate in private locations near the periphery of their home range. Cats seek a secluded area, say a patch of clean, dry earth in a quiet spot behind some shrubs and bushes, dig a hole and relieve themselves. They then carefully cover their excreta. This explains why indoor-only cats are generally reluctant to use dirty litter trays or those placed in convenient but public areas. Choosing trays of suitable design, litters that our cats like, placing his tray somewhere private and keeping it clean should help avoid unnecessary stress and the possibility of house soiling.

TYPICAL FELINE DAILY ACTIVITIES

Glancing at a list of everyday feline activities can help you create a truly cat-friendly environment and avoid the possibility of your cat developing behavioural traits that will put you at odds with each other.

Activity	Indoor facility	Outdoor facility
Running	Hallway or stairs	Garden or yard
Jumping	Table tops; into boxes or sheets of newspaper on the floor	Fences; trees
Climbing	Activity centres; furniture	Fences; trees; pergolas
Exploring	Open cupboards and drawers; boxes; under beds, behind sofas	Ponds (covered); trees; shrubs
Observing	Window sills; stairs; tops of furniture; activity centres	Sheds; fences; walls; under bushes
Resting and grooming	Cat and human beds; sofas and chairs; under furniture or radiators	Safe sunny spots; inside greenhouses, sheds; under dense foliage
Marking	Scratching posts or pads near entrances and exits to the home and rooms that are important to him	Trees; fence posts; wooden garden furniture
Playing	Anywhere with lots of toys	Plenty of natural stimulation
Eating	Where he feels safe; not noisy or busy or near litter tray	
Drinking	Variety of places not near food or litter tray	Natural puddles or pools
Hiding	Under beds; inside cupboards; boxes	Dense shrubs; behind large flowerpots; under garden furniture
Eliminating	Litter trays in a quiet, private location	Sheltered private area with dry patch of earth, near the periphery of territory, such as side or bottom of garden
Socializing	Wherever and whenever he chooses!	

Happy relationships

Some cats have relaxed, happy relationships with their human companions, others may keep their distance. If you try too hard to 'make friends', a cat is likely to find such overtures inadvertently pressurizing and stressful.

Past experiences – a cat may be shy or nervous for many reasons, perhaps because he was not well socialized as a kitten, is of a naturally nervous disposition or has had unfortunate past experiences of people or other animals.

Playing it cool – the best approach is always to allow cats to take the initiative. Avoid eye contact, remain at a distance and make yourself small by crouching down and side-on to an inhibited cat. When he's more confident hold out your hand while crouching

A LOVING RELATIONSHIP

Don't feel hurt if your cat is not a lap cat. Many cats show their affection by just being around. These pets may just greet their people with a chirrup and a quick rub around their legs, then be on their way, checking in from time to time. Others enjoy cuddles, while many prefer communal games, especially when you're doing something else like working at the computer or reading the paper. The best human-cat relationships are the ones where people accept their cats for what they are and choose a style of love and attention that suits their pet.

WHAT NOT TO DO

Talk too loudly

Become impatient

Stroke a cat too soon or too hard

Pat his head or hindquarters

Pull his tail

Overdo it – keep sessions short but sweet

Pick him up until he is completely familiar and comfortable with you

down and not looking directly at him. Let him sniff you for a while and if he seems relaxed perhaps give him a quick but very gentle stroke. Aim for under his chin as reaching out to touch a cat's head may feel threatening to him.

Softly, softly – once he starts to approach never make the mistake of rushing over to impose attention but always respond, at first in a low-key, hands-off manner, to any overture a cat makes. Talk in a quiet monotonous tone and use fishing-rod or thrown toys as 'an invitation to interact'.

Slowly does it – gradually building up contact at the rate he indicates he is comfortable with may be a bit frustrating but it is likely to ultimately produce the sort of bond cat lovers find so rewarding. Observe your cat's body language and behaviour to judge how he's feeling.

The right cat for you

Once you have decided to share your home with a cat, some important decisions have to be made. Your lifestyle and the age, personality, breed and health of your pet should all be considered. If you are prepared from the start, you will be well on the way to providing a happy home for your cat.

Choosing your cat

Before you head off to acquire your cat, it is worth asking yourself questions about the age and type of cat you would prefer, what kind of lifestyle he will have and how he will fit into the existing family.

THINGS TO CONSIDER

Will a kitten or adult suit you best?
(See opposite)

Is pedigree important or would you love an 'ordinary' domestic cat? (See pages 48–51)

Longhaired or shorthaired?

Male or female?

What sort of lifestyle do you want your cat to have: one where he goes outside or lives entirely indoors? (See pages 52–53)

Do you have other pets, particularly cats, to consider? (See pages 82–87)

Are you often away from home?

Can you afford regular health care?

Kitten – a healthy kitten is a ball of fun! It is also quite hard work and requires a significant input in terms of time, effort and entertainment from everyone, if he is to become a well-balanced and satisfying pet. Some people make the mistake of thinking that by getting two kittens their role will be diminished, as the youngsters will entertain each other. To some extent this may be true but providing for them and clearing up is still

KEY QUESTION

Are male or female cats better as a family pet?

Provided they are well socialized, especially with children, at an early age and are not overly fearful in nature, both male and female cats make great pets. Neutering is also advisable to prevent problems, such as roaming and fighting, associated with sexual activity.

double the work – and it always pays to factor the time wasted watching kitten antics into your calculations!

Adult cat – especially if he is re-homed directly from someone whose circumstances force them to relinquish him, has the advantage that his looks and personality are already established. Settling him should be relatively straightforward.

Golden oldie – improvements in nutrition and health care mean that many more cats live well into their teens and, through no fault of their own, more 'golden oldies' need loving new homes. Many people are understandably attracted to giving deserving old timers a happy retirement and often find that they make rewarding companions. However, expectations of older cats must always be realistic.

Commoner or aristocat?

The common 'moggy' is as delightful and varied as any pedigree, but if you would prefer the refined looks of a pure-bred cat then there are plenty to choose from. Pedigree cats have distinct character traits because they are bred from a relatively small gene pool. The following is a brief guide to the main pedigree breeds; for more detailed information, research further in books or on the Internet.

Abyssinian – independent, intelligent, playful cats that like learning tricks, these elegant creatures are not good in multi-cat homes but make playful companions for people who are not daunted by adventurous curiosity.

Bengal – these cats are bright, confident, easily bored and generally (when well socialized) drawn to human company. They are not pets for people who are rarely at home.

Birman – these sensitive, pleasant-natured lap cats make great playmates for people who have plenty of time for them, because they can be distressed by solitude. Being longhaired they also require regular grooming, which is not always ideal for busy owners. They are a good choice for multi-cat groups and indoor lifestyles.

British Shorthair – these are placid cats that enjoy humans; they make good pets, although owners must carefully control food intake and encourage exercise.

Burmese – not for the faint-hearted, this intelligent, inquisitive breed tends towards vocalization and attention seeking if not sufficiently stimulated and entertained by owners to whom they generally bond firmly, making great companions.

Forest Cats – Norwegian and Siberian Forest Cats enjoy the great outdoors and need space to roam. They moult heavily annually and the former in particular are considered sociable and quite easy-going.

Maine Coon – another longhaired type, although the coat is relatively easily managed, these cats are large in every way. Their independent, adventurous natures and lively intelligence ensure they quickly pick up tricks making them entertaining for those who do not want lap cats. These hardy creatures are not, however, predisposed to live happily in small homes, multi-cat households, as indoor-only pets or in areas with high feline populations.

Commoner or aristocat?   **49**

Ocicat – another bright, reactive, easily bored breed that doesn't cope well with isolation from owners.

Persian – usually gentle and playful, if inclined to clumsiness, Persians are not ideal for anyone who is not blessed with the time to keep them well groomed.

Ragdoll – gentle and tolerant when well socialized, Ragdolls tend to make good 'cuddle cats' and satisfying companions with an aptitude for training and relatively low exercise requirements but individuals vary. Be prepared for the exception that proves the rule. Coat care may, however, be an issue with this longhaired breed.

Rex Cats (Devon and Cornish) – have short wavy hair, are generally sociable with people and are said to enjoy the company of other pets, including suitably chosen cats.

Russian Blue – shy with a need for stability and gentle handling, these bright cats make rewarding companions for people with whom they are familiar, although they may not be the ideal choice for a household with children.

Siamese – full of character and invariably willing to use their voices to get what they want, these cheerful cats enjoy company and fare badly when left alone for long periods of time. With a tendency towards attention seeking, they require plenty of mental stimulation to prevent them from being bored, becoming difficult to live with and developing behavioural problems.

Turkish Van – many people value the intelligence and willingness to participate in training that comes with their Turkish Vans, a breed well known for a natural love of water and sociability when well introduced to the human world.

Lifestyle choices

When choosing a cat, you need to think carefully about his lifestyle. You should consider the effect of other pets, whether he will live solely indoors or be able to roam, whether he will be the only feline or part of a multi-cat household.

Commitment – as with any pet, you should ensure that you have the time to play with and care for your cat, being there to feed him regularly and carry out health checks and routine grooming.

KEY TIPS FOR MULTI-CAT HOUSEHOLDS

Identify the separate feline social groups within your home: related cats may choose to spend time together and will form one small colony; loners should be treated as a separate social group, with separate facilities

Provide sufficient separate toileting, feeding and sleeping facilities plus individual hiding places for each group or colony

Keep toys in separate places

Recognize signs of stress (see pages 146–147)

Other pets – cats and dogs come from different species and this is an important point to consider when introducing a new cat into a home with an established dog in residence. Some breeds, Terriers in particular, instinctively chase, and sometimes kill, smaller pets. However, there are other dog breeds that will tolerate a feline companion. (See also pages 86–87.)

Indoor or outdoor – your domestic circumstances may dictate that your cat has to live his life inside (you may live in an apartment, for example), or you may choose to keep your pet safe indoors rather than introduce him to the established neighbourhood cats or the dangers of a busy road. Outside cats have the advantage of indulging their natural behaviour to explore and establish a wider territory.

Local population – it is worth bearing in mind that when there are lots of cats in your

vicinity, your property may well have been 'carved up' between feline neighbours. An adult newcomer may have more difficulty than a youngster establishing his ownership rights, although the physical safety of any incoming cat can never be guaranteed in areas with a high feline population and outdoor introductions should be supervised.

Multi-cat households – if you are considering getting two or more cats, bear in mind that things should be handled sensitively. Unrelated cats forced to live together and play 'happy families' is frequently a recipe for chronic, low-grade stress. See the box opposite for tips on helping your cats to get along.

Where to get your cat

Now that you know a cat is right for you, your next decision is where to find one. Some lovely cats turn up looking for homes through family or friends, you can purchase a pedigree from reputable breeders, buy a kitten or re-home a rescue cat. Be prepared to ask the right questions so that you are sure your chosen cat is right for you.

A good start – when buying a kitten, you should also meet his mother; her personality will in some part be reflected in her kittens. Look for a bright-eyed, sociable kitten that is happy to approach you and be handled, and check that his living quarters look clean and well maintained. If the cats on offer look ill kept, unhealthy or overly fearful you may

This is not always easy. Look underneath the tail. A female's vulva should appear as a small opening close to the anus while the male has a greater distance and the potential testicles interposed between his anus and penis.

have to turn them down. Such sad little creatures may make good pets but a poor start in life does not generally bode well for the future.

Rescue cats – although their previous history may be unknown, always ask staff at the rescue centre as many questions as possible so that you have some idea of a cat's personality and health status before making a decision.

Family cats – unless cats are exposed to people of both sexes who represent a variety of age groups, they usually have difficulty fitting into a busy family home. Living with young children can be problematic for cats that didn't meet them when they, too, were immature and their behaviour was more malleable. Making informed and realistic choices is essential to avoid disappointment.

ASKING THE RIGHT QUESTIONS

Wherever you go for your cat and whatever his type, ask some searching questions and make telephone enquiries before you arrive. You want a well-socialized cat and sufficient information to identify any risk factors that could make him difficult to live with.

What vaccinations and parasite control measures have been undertaken?

Any history of previous illness, accidents or medical treatment?

Were the kittens raised indoors in an area where they can see and hear all sorts of domestic activities, such as a kitchen?

Did an adult cat previously live in a family home? If you want a household pet at ease in your home, choose a cat that is familiar with an indoor environment

Who handled the kittens or what sort of household is an older cat used to?

Has the kitten or cat been introduced to children?

Bringing your cat home

Introducing your cat to his new home needs preparation, time and patience. Have everything ready for his arrival and prepare a special sanctuary where he can settle in at his own pace. His first impressions will be lasting ones – make sure that they're positive and handled sensitively and the household will be one of peace and harmony.

Planning ahead

You are always more likely to forge successful relationships with a new cat if you prepare well ahead of his arrival. Cats are not all the same and cats of different ages have different needs.

Equipment – have all the basic essentials (see pages 60–61) ready before you collect your cat. However, don't spend a lot of money before you really know your new cat, his preferences and quirks of nature, and bear in mind that kittens may quickly outgrow climbing frames, beds or litter trays.

Travel carrier – a prepared travel basket, box or carrier (with comfy bedding and a layer of newspaper) is essential for transporting your cat safely and with the minimum of stress.

Veterinary care – research a cat-friendly veterinary practice (see pages 94–95) and make an appointment in the first few days of bringing your cat home. In this way you can ensure that he is fit and healthy, and discuss any routine care issues with a professional.

Food – discuss his dietary requirements with his previous carers. It is advisable to use the same type or brand of food that he is used to for the transitional period.

Other pets – think carefully about where your new cat is going to reside for the first few days in his new home, and make sure that established pets are restrained or shut out until proper introductions can be made (see pages 82–87).

Timing – plan the best time to bring your new cat home. Ideally, avoid times of upheaval or change, such as moving house, guests staying or redecorating. Devote some time to helping your pet settle, taking time off work if necessary.

Essential equipment

Whatever your new cat's age, you need to cater for his basic needs, particularly at first when you keep him inside to settle and bond before allowing him outdoors. Adequate feeding, sleeping and toileting facilities are the basic requirements and bear in mind these may change as he grows and ages.

Beds – provide two alternatives as cats should always have a choice, it helps give them a sense of control, which is important for emotional equilibrium. Offering different designs also helps you establish a new pet's preferences, while providing at least one igloo-style bed allows your cat a cosy refuge as well as somewhere to sleep.

Blankets – you will need several. Never wash everything at the same time, especially initially when maintaining some reassuring scent helps provide essential comfort. Having a blanket in each bed and another one or two in comfortable locations means all your cat's familiar 'landmarks' do not have to disappear at once on washday.

Food and water dishes – at least two sets are essential so you can keep them clean.

Litter trays and litter – choose between an open or covered design and ensure that the tray is large enough. Cats can be particular about the type of litter used, and you may have to experiment to find the one that he prefers (see also pages 74–75).

KEY TIPS

If you use recycled equipment, clean it thoroughly to remove any intimidating lingering scents and check it is appropriate for the cat that's now coming to stay. Kittens and elderly or disabled cats, for example, can find it difficult to climb into high-sided litter trays and beds.

If you are purchasing a new carrier it is often advisable to buy one that is constructed in two halves, so by removing the top your cat has a comfortable bed that will acquire his individual scent and be reassuring when he travels.

Collars and ID tags – even if your cat will be a 'stay at home' pet it is sensible to give him some visible means of identification in addition to his microchip (see page 99), just in case he accidentally escapes. All cat collars should have a failsafe device that allows quick and safe release if the collar gets snagged on something.

Grooming equipment – a variety of brushes, combs, grooming gloves and rubber pad-like brushes are available. Choose soft, gentle brushes for kittens or frail elderly cats and ask advice of previous carers if you are taking on an adult cat.

Carrier – each cat should have his own carrier with which you condition positive associations so that travelling in it does not become unnecessarily distressing for him. It should be secure and also easily dismantled so that it can be thoroughly cleaned.

Your cat's toy box

To prevent boredom, your cat needs to indulge his instincts to chase. Toys are an essential item for all cats, providing entertainment and a great opportunity for you to interact and bond with your feline companion.

KEY QUESTION

When playing with my cat, if often ends with him grabbing my hand rather than the toy. How can I prevent this?

The best toys for interactive play are those that you throw or things you can safely dangle, such as fishing rods or feathers on sticks. When cats get very excited, they tend to lash out and latch on to anything that moves, so hands invariably become a target. 'Read' your cat well and stop playing before he reaches fever pitch, throwing the toy down to distract his attention away from you.

Interactive toys – fishing-rod toys and sticks with feathers attached make an ideal way to play with cats without imposing close contact attention, which can be intimidating even though that is not our intention.

Homemade toys – cardboard boxes and tubes, balls of tissue paper, paper carrier bags (remove their handles), are good examples of 'recycled toys' that can provide hours of feline amusement for nothing. Ensure they are safe with no sharp projections or easily detachable fabric or plastic bits that could be swallowed.

Simple toys – cats prefer 'prey-like' items – successful toys are small, light and easily batted about with paws. If they make a noise that is not loud and intimidating so much the better, which is why those that squeak or have bells are often so prized by their feline owners.

Scratching posts and pads – ensure these are stable and appropriate in height to your cat's size and age. Your cat should be able to stretch completely while scratching. Scratch pads are also available.

Climbing frames – are excellent for giving cats somewhere high enough to sit safely out of the way and watch what's going. Kittens need smaller versions they can safely climb but for an adult cat a climbing frame must be tall with platforms, tubes and enclosed boxes that he will still be able to fit into when he matures.

Preparing accommodation

Whatever his age and background, it is crucial that your cat has a quiet sanctuary when he first arrives so that he can gradually settle before starting to explore his new environment.

A quiet spot – cats need a safe, peaceful spot to rest and observe from. Ideally your newcomer should be given his own room but this isn't always practical, so choose the corner of the quietest room for example.

Hidden sanctuary – to help him feel secure, provide a covered area with bedding: this could be an igloo-type bed, a cardboard box or an indoor kennel or cage. However, it should never be put on the floor if you have small children and other pets, particularly dogs, which will all be in direct contact with him on the same level, which he will find intimidating. Cats retreat to high places when they are unnerved, so positioning his sanctuary in an elevated spot, on a table or cupboard, for example, will feel natural.

Safe retreat – another advantage of a cage, which should always be viewed as a positive refuge by its inhabitant, is that a tiny kitten can be safely enclosed when there is a lot going on, thus avoiding potential accidents during busy domestic activities. They do however remain in 'social' contact with everyone as they can see, hear and smell all that is happening.

SANCTUARY CHECKLIST

Whether your cat has a room or large cage to himself he needs:

His bed

A cardboard box – for a refuge and a raised perch on which your cat can sit and survey his new surroundings when he feels more confident

Litter tray – as private and as far from his food and bed as possible

Food and drink bowls

Toys

Scratching facilities

Extra measures

When thinking ahead, include the following in your preparations to reduce the risk of accidents and some of the stress-related problems commonly referred to behaviour counsellors.

Stress relief – commercial pheromone preparations have been developed that help reduce cats' stress levels (see box, opposite). They are available either as a plug-in diffuser or as a spray version, which is useful for travelling. Carefully follow the manufacturer's instructions and install the electrical diffuser in time for it to be effective.

Scent sensitivity – if the newcomer's sanctuary is already used by other pets, close the door and reduce the potentially stressful impact of their scent by washing items they have slept on, for example.

Extra facilities – acquire any additional facilities, such as litter trays, beds and feeding bowls, you need to enable your pets to keep out of each other's way if you have a multi-cat home.

Canine control – think of installing a dog gate as a means of controlling the canine contingent but allowing cats to escape over the top, if adult and agile, or through the bars, if small and sinuous. This may be a useful addition to your home at least in the early stages when relationships are being cultivated under your careful supervision.

Safety first – build safety measures into your everyday routines. Cats, particularly inquisitive kittens or bored indoor-only adults, are always at risk of being involved in domestic accidents.

- Hide or tape down electric cables
- Close doors on computer rooms
- Keep clothes washer and dryer doors shut and check before you switch them on
- Buy a cover for your electric hob
- Remove poisonous houseplants or flowers (see box above and page 91)
- Tidy away sewing paraphernalia, especially needles with threads attached – these can be lethal
- Fit childproof locks to cupboard doors that could allow a cat to access poisonous substances, wriggle under floorboards or into other risky places
- Fit screens to windows and enclose balconies to prevent cats falling from upper floors, especially in apartments or homes with loft rooms. Never take chances with your cat's safety

TOXIC HOUSEPLANTS

Amaryllis
Castor oil plant
Chrysanthemum
Cyclamen
Dumb cane
Devil's Ivy
Elephant's Ear
Ferns
Hyacinths
Lilies
Poinsettia
Umbrella plant
Zebra plant

PHEROMONOTHERAPY

This is a recently coined term, which refers to the use of synthetically derived, commercially available feline facial scents – the reassuring scents that cats release when rubbing their faces around furniture or our legs, that makes them feel calm and relaxed. It can be used to relieve stress associated with moving house, travel, trips to the vets and new arrivals in the home.

Handling your cat

How much handling any cat desires depends upon his age, temperament and history. Young kittens are fragile and easily hurt and must be handled carefully. Early painful experiences can give rise to negative associations, and sometimes make cats resentful of handling altogether.

Close contact – on leaving their old home youngsters have lost everything that is reassuringly normal – contact with mother, siblings, carers, familiar sounds and smells – and they have no previous experience on which to draw to help deal with the inevitable stress of changing homes. Physical contact is likely to be more important to them but overwhelming kittens at this stage can be counterproductive.

Gentle handling – the knack with any cat is to support his weight with one hand or arm while restraining him with the other to prevent him falling but to avoid squeezing too tightly. Place one hand under the hindquarters and one behind the front legs.

Respect – whatever a cat's age, the general rule of 'allowing him to come to you' always holds true. Following each individual cat's lead and remembering that, even if he is

THE TRIP HOME

When you collect your new cat, make sure you approach calmly and confidently, make yourself small and non-intimidating by crouching down, side-on to him and avoid eye contact, just talk quietly. Getting the previous carer to tell you how he likes to be handled and watching them lift him into his carrier when you collect him is useful, especially if he is nervous. You then avoid being associated with his aroused emotional state.

very sociable and inclined to spend time with you, the newcomer may not be a lap cat is a golden rule we should respect.

Time limits – keeping actual contact short and sweet pays dividends. It often goes against our natural inclinations but even with people to whom cats are well bonded, the

feline way is to check in often but remain somewhat remote – a quick rub against our legs, stroke in return, a brief chat and perhaps a short game and they're off.

Accepting this rather than 'trapping' a cat into a cuddle is a better way to encourage him to spend time with you, even if he just falls asleep beside, rather than on, you.

Settling in

The first night, and following few days, will be exciting for you and relatively disturbing and stressful for your new cat. Help him to settle and give him some space to familiarize himself with his new home and he will soon be part of the family.

The first night – introduce your cat to his sanctuary (see pages 64–65) so that he can get his bearings, then make sure he has all he needs and is comfortable before leaving him to settle – tough but wise and kind.

Including a well-wrapped hot water bottle or special pet heat pad (again well covered for safety) for youngsters or elderly cats, and something from his previous home that smells familiar and reassuring will help.

Check in – from time to time to make sure he is OK but don't hassle, as constant intrusion will be stressful. If your new pet chooses to approach you that is great but don't push the pace – remember the upheaval he has experienced and the feline need for control.

KEY QUESTION

I've heard that putting butter on a cat's paws helps them to settle. Is this true?

Probably not, but giving a cat something to concentrate on when they are in a new environment is often helpful in taking their minds off things. A better way to help cats settle is to ensure they have a safe, secure sanctuary with lots of hiding places, some items with their own familiar scents on and a commercial pheromone diffuser plugged in (see page 67). Then allow them enough time to gain confidence before introducing them to their new home and other household residents.

SHARED SPACE

It is a matter of personal choice but, if their new cats have been wormed and treated for external parasites, some people allow them to settle initially in someone's bedroom so they do not feel isolated.

Hideaways – some cats, in fact, simply hide for two or three days but their new owners are reassured by the fact that when everyone has gone to bed the food is eaten, litter trays used and toys moved about – all signs that everything is well but the newcomer is still feeling shy and a little overwhelmed. He'll emerge when he's ready.

Routine care

There are a number of essential daily chores attached to cat ownership. Attending to them diligently helps us keep pets healthy and avoid any potential problems, such as house soiling due to poor litter tray hygiene.

Daily 'to do' list – you will soon establish a daily routine for looking after your cat and he will probably quickly come to expect his meals at regular times, appearing to remind you if you are a little late!

- Food bowls should be washed at least once a day, although after each meal is best
- Water dishes should also be cleaned every day and then refilled with fresh water of the type each cat prefers (see pages 136–137)
- Toys should be checked to make sure they are not soiled or damaged in a way that could make them potentially dangerous
- Introduce a new plaything most days and rotate them to maintain novelty, especially while your cat is kept indoors
- Attend to litter trays, clearing them of soiled litter and faeces (see box, right)
- Check your cat over thoroughly to ensure he has no signs of illness or infection (see pages 112–113)

- Groom your longhaired cat. Add it to your list today if your pet is shorthaired and has not been brushed for a few days, especially if he is moulting because it is hot or you have turned your household heating up during cold weather. Ingested excess fur can form hairballs, which cause vomiting, constipation and occasionally intestinal obstruction that requires surgery

LITTER TRAY HYGIENE

Cats are generally fastidious and find it stressful to use a dirty litter tray. As soon as the litter has been used, the soiled patch should be scooped out. The tray should be well cleaned, dried and all the litter replaced at least twice weekly – more frequently if a cat has diarrhoea or an illness, which causes excess urination. Do not use highly scented or lemon-scented cleaners, as these may be off-putting to your cat.

- Wash your cat's bedding regularly, and avoid using biological cleaners and detergents, which can cause skin irritation and reactions.
- Tidy away any possessions that could harm you cat, such as your sewing kit, hobby materials or discarded food wrappers. Your cat may be injured or become ill as a result of investigating, playing with or swallowing such items
- Periodically check your home and garden for potentially hazardous changes and deal with them straight away. For instance, cupboard doors that don't close and allow your cat access to toxic substances should be repaired immediately

Litter trays

As cats naturally seek out private locations near the periphery of their home range for latrine sites, we need to show sensitivity not only in what we provide but also where we put their litter trays. Avoid placing them in busy areas of your home.

Design – covered trays have the advantage of privacy but some cats find 'entering the tunnel' intimidating, especially if having to pass through a flap in the door is a novel experience. Off-putting fumes can also build up if they are not cleaned properly and frequently. Provide different designs to see what your cat prefers.

Size – the tray should be large enough for a cat to turn around comfortably in and the sides low enough to climb over.

Litter type – clumping litters, which are relatively similar in appearance to the sand that their ancestors encountered, are generally preferred by most cats.

Litter depth – cats need a reasonable depth of litter in order to bury their excrement to their satisfaction, so it is a false economy to skimp on the amount. Ensure the litter is at least 4 cm (1½ in) deep.

Additions – such as the use of strong disinfectants, scented deodorizers and plastic tray liners, can put some cats off using their trays and lead to house soiling.

Choice – the number and location of litter trays is crucial. There should be one tray per cat plus an extra and they should be split up so that each feline social group has its own latrine site within its own space.

Housetraining

Most adult cats are already housetrained and it is relatively straightforward to train a kitten to use a litter tray and, when he is mature enough, he will use the natural facilities if he is allowed outdoors.

Training your kitten – most kittens will have been litter-trained before they are re-homed by copying their mother. You can ensure continued success by:

- Making sure your kitten knows where his litter tray lives
- Always having easy access to a tray; you may need to have a number of trays if your property is large
- Recognizing the signs when he needs to go (see box, above)
- Reward him for using the tray with a small titbit but take care not to interfere too much or you may put him off his tray

'Accidents' – there can be several behavioural causes for house soiling in older cats, often linked to cats that are feeling threatened or nervous. For example, a nervous cat is unlikely to go through a busy, noisy kitchen or past other pets to get to his latrine, especially when he has recently arrived. Urinary infections or other health issues may also play a part, so it is important to monitor your cat carefully to rule these out (see page 113). **Never** punish your cat – work out why the problem occurred and rectify matters.

CLEANING UP

Do not use ammonia-based disinfectants – they simulate urine and encourage further use of the affected area

Use disposable paper towels

Where surfaces will allow, clean with a 10% solution of biological detergent

Rinse well with water and dry

Repeat several times

Also treat underlays and floors if urine has soaked through

Using the cat flap –
kittens and adult cats that are not used to cat flaps may be frightened of the noise and sensation of pushing against the flap. Start by gently pushing it to and fro when your cat is doing something pleasurable, like eating, nearby. Once he is used to the noise, use treats to reward his interest or to tempt him through with the flap fastened open. Be patient, never force him but make it worth his while to be adventurous with the flap and he should gradually become accustomed to it.

House rules

Young cats are learning all the time. They do so by exploring, watching others and because we reward what they do. While it is not possible to train a cat in a similar way to a dog, they will soon understand what is acceptable behaviour in their new home.

Realistic rules – make sure that the rules are appropriate in view of his age, personality, history, species behaviour and circumstances.

YOUR CAT'S EXPECTATIONS

Allow your cat the freedom to express himself by providing lots of dedicated, appropriately located facilities on which to:

Exercise – clear space to run or allow access to outside

Practise his climbing skills – access to tops of wardrobes, kitchen units

Seek elevated, enclosed refuges in which to hide – access to a cupboard

Get up high to survey what's going on – clear a shelf

Sharpen his claws and mark his territory – provide scratching posts and pads

Consistent – be consistent with what is and what is not acceptable, and make sure that other household members, especially children, apply the same rules in the same way. For example, letting your cat sleep on your bed but not on your children's beds is unrealistic.

Enforcing – don't use punishment or aversive methods, such as shouting, to enforce the rules. Instead, think ahead to try to prevent problems, or use a gentle 'no' and distract your cat into doing what you want with a toy and reward compliance.

Expectations – always try to reduce the motivation for an inappropriate behaviour rather than dealing with it by having realistic expectations and catering well for your pet. For example, cats climb, so if you fail to give them appropriate equipment on which to undertake their adventures, they are likely to use your best curtains or knock ornaments off an available shelf.

Exploring the home

A common mistake that people make is to underestimate the time it can take a new cat to settle. Some, of course, are confident and ready to explore within a couple of days but others take weeks to adjust to their changed circumstances. Respecting this is more likely to produce a well-adjusted pet at ease in his new surroundings.

Day by day – once your cat has settled well in his sanctuary and is happy to meet you, start leaving the door open for him to explore further afield if he wants to. Never force him out and never prevent him from running back if he's spooked. If he's bold, especially if yours is a quiet household, he may well feel confident to tackle a new room each day until he is well acquainted with the whole house. If his is a more timid disposition, he may well take several weeks to pluck up the courage to come out of his refuge.

Appliances – be especially sensitive when introducing your newcomer to all the household appliances. Smoke and burglar alarms, vacuum cleaners and other electrical equipment often make highly artificial noises. Protect your cat by creating his refuge as far away as possible from them. Never force him into close contact and if he tries to run out of the room, let him or gently move him to a distant area of the house as a temporary measure.

Escape routes – make sure his sanctuary is available for him to rush back into if he is unnerved or suddenly feels insecure.

NEW FACES

Carefully conduct introductions to the other individuals with whom a cat will live. At first it is advisable to fuss a cat as little as possible and to ensure his main carer handles him. Then work slowly through introducing everyone in the household, supervising children carefully (see page 84). Initial contact is best kept to hands-off invitations to play, which leave the cat very much in charge of the situation.

Observation posts –
wherever he is ensure there
are easily accessed, elevated
areas and dark hiding places
for him to use to sit quietly
and observe before he goes
any further – cardboard boxes
can be useful.

Scent reassurance – use
commercial pheromone products
(see page 67) to create a
reassuring environment just ahead
of his explorations to reduce the
impact of 'the new'.

Meeting other cats at home

Trying to force a new cat to make relationships with existing household cats – who are not part of his social world as far as any of the cats are concerned – is often disastrous. Accepting that your cats may at best become distant but amicable, sharing the same home but giving each other a wide berth, is wise.

First introductions – don't rush into releasing your new cat into a multi-cat household, or into the main domain of a single cat. Follow the guidelines on pages 70–71 to keep the new cat in a safe, secure room and gradually introduce the cats, using the simple steps given below.

Signs of distress – be especially sensitive to both the new cat and the existing cat's behaviour. This is especially important as many cats in multi-cat homes develop stress-related problems that are not identified because of the natural feline tendency to retreat when under pressure.

Emotional impact – resident cats also need TLC (tender loving care) and probably extra de-stressing facilities as well.

Own space – once the newcomer has settled, and you have provided extra hiding places and high refuges so all your cats have escape areas, gradually allow them to adjust to each other. Cats always cope better when they can see, hear and eventually meet each other without undue pressure.

SIMPLE STEPS TO SUCCESS

Before face-to-face meetings, use the cats' own scents for initial introductions. This naturally happens when they rub against your legs and you go between your feline groups. When they smell each other's scent make it positive with very special titbits.

1 Wipe each cat with a separate cloth and present this with food or play, gauging his reactions. Don't move forward with introductions until any negative responses have died down.

2 Then mix the cloths in a bag to produce a colony scent. Repeat the process allowing time for adjustment before the newcomer is let out to explore, when the other cats are outside or asleep in a closed room.

3 When your cats eventually meet, take care not to force things. Have lots of toys with which to distract them from each other and create a positive atmosphere.

4 Don't forget longer term to maintain separate feeding and living areas based on the individual feline social groups (see page 53), otherwise you could unwittingly cause distress.

Introducing other pets

Unless cats are appropriately exposed to pets of other species during the socialization period they may never be comfortable (or safe) in close proximity to them. However, bringing kittens up to view other household pets as part of their own social group is sometimes possible.

CATS AND CHILDREN

Many believe having pets encourages responsibility and empathy in children. However, while involving youngsters in their cat's everyday management, parents must accept ultimate responsibility for his care and supervise interactions. Overenthusiastic children can sometimes be inadvertently rough and intimidating.

Supervision – is key to making a success of the enterprise and cats should never be left unattended with any pet from a prey species, no matter how well bonded they may appear! Equally, when introducing a resident dog, never leave them unsupervised (see pages 86–87).

Refuges – make sure that your cat can quickly escape to his sanctuary or an elevated observation post. It is important to maintain a degree of separation between pets, especially where dogs are concerned, so that the cat feels safe in one part of the home.

Smaller pets – it's important to be aware that for many small furries, such as hamsters, even the scent of a cat is bad news. Our felines are predators so we must take care not to allow them access to smaller pets from prey species, as well as ensuring we don't pet our cat, then handle our tiny companions.

KEY QUESTION

If I get a kitten and a puppy at the same time, will they be friends?

Quite possibly the answer is 'yes' so long as you make a good choice of well-socialized pets and carefully conducted introductions. Choose a kitten whose parents are bold and sociable, and a dog with confident but steady parents, preferably of a breed that was not developed to chase small furry creatures.

Cats and dogs – the breed, temperament and age of our dogs and the degree of control we are able to exert are important issues when introducing them to a new feline companion. A poorly trained dog should never be introduced to a cat and exceptional care should be taken with Terriers.

Obedience – brush up on your dog's obedience training before the new cat arrives. You need to feel confident that he will obey commands to leave and sit.

Escape routes – a frightened cat that is unable to escape is unlikely to tolerate a dog again, let alone become friends with it. The secrets of success are:

- Elevated refuges
- Dog gates over which cats can escape
- A light lead that the dog wears for control if/when necessary
- An endless supply of favourite treats or an exciting new toy with which to distract your pooch
- Use pheromones to create a calm environment (see page 67)

Closer contact – repeat the sessions over several days, weeks or months depending upon your pet's reactions. When the time is right, allow your cat to approach the dog but not the other way around. Never leave your pets together unsupervised until you are sure they are well adjusted and your cat can easily get away.

QUICK GUIDE TO HAPPY CANINE INTRODUCTIONS

Keep introduction sessions short and sweet. Control everything and ensure the dog learns that it is more worth his while not to chase the cat than to do so.

1 Do not force your cat into contact with your dog but when you think that the time is right sit quietly in a room with your dog beside you as far from the doorway as possible.

2 Keep hold of the lead but let it hang loose or put your foot on the end so you can stop the dog chasing the cat. Have your supply of treats or new toy nearby and make your dog sit with his attention focused on you.

3 If you have someone to help you, get them to play with the cat and gradually move towards the room but not too close. If not, choose a time when the cat is likely to approach – lay a trail of dry food perhaps to encourage him – wait until he comes in.

4 Reward your dog for not taking any notice or just sitting quietly and let the cat hop up onto a shelf, climbing frame or the back of the sofa.

5 End the session before your dog gets excited or if the cat becomes stressed or frightened.

The great outdoors

When equipping their homes in preparation for a new feline arrival, many owners overlook their outdoor space. Be sure to provide a cat-friendly environment where they can indulge their natural behaviours.

Latrine areas – if possible, dig over an appropriate private area to create a pit, then fill it with gravel and add some good-quality play sand to the soil. Clean the latrine regularly to prevent your cat from searching out freshly dug areas in your prize flower borders.

Hiding places – as for indoors, cats love to have places to hide and observe from. These can be provided by outdoor kennels, large plant pots and tubs arranged near exit doors or the cat flap. Shrubs and trees provide shady resting spots and elevated positions.

Climbing – shed or garage roofs, fences, playhouses and climbing frames all provide platforms that allow cats to get up as high as possible to observe what's happening around them.

Territorial markers – your cat will need somewhere to leave a marking scent and visual marker to declare his territory boundaries. Tall, stable, soft wooden objects, untreated with chemicals, such as fence posts, are ideal. Provide dedicated scratching posts to protect garden furniture and trees.

DESIRABLE EXTRAS

Hang mobiles, glittery ornaments and/or feed birds (very high and out of reach) for interest and entertainment

Grow cat grass and catmint

Block easy access points for other cats using flimsy trellis

Never encourage neighbouring cats into your garden

Exploring the world outside

Once your cat is really well bonded to you and settled in his new home (allow three to four weeks), you can introduce him to the outside world. Prepare your garden, making sure it fulfils his requirements (see pages 88–89), and is safe and secure.

OUTDOOR HAZARDS

Check your garden for the following before allowing your cat outside:

Many chemicals used to treat wooden garden items are toxic

Antifreeze is a relatively common cause of poisoning in cats, so ensure garages or sheds are well cat-proofed

A range of plants are toxic if ingested (see box opposite and page 67)

Deep ponds and swimming pools can be hazardous

Keep plastic rubbish bags securely contained to avoid injury from jagged food tins or cooked bones that shatter when eaten

ID tag – before he ventures outside make sure he has a safe collar fitted with an ID tag with your contact details, just in case he can't find his way back. Microchipping is an additional method of identification (see page 99), but not as instantly accessible.

All clear – ensure the locality is quiet and you are around to supervise and possibly intervene if neighbourhood cats pay a visit.

Homing incentives – choose a time when he is hungry and most likely to come back when you call him in for food.

Freedom – for the first few times, leave his exit/entrance door open and allow him to decide if he wants to go out at all and how far he ventures – he may keep popping out and then darting back quickly until he feels confident. Strategically located hiding places for him to use should also help him to feel more at ease.

Confidence – gradually build up the time he spends outside once you are sure he will return safely and knows the area as well outdoors as he does at home. Once he is familiar with the cat flap (see page 77), he can be fully independent.

Outdoor pen – this can be an asset in areas with busy roads, a very high feline population, neighbours who are not feline-friendly or local laws that prohibit cats outdoors. The pen should be secure, with a cat flap to give your cat the choice of using it, and well furnished with shade and shelter.

TOXIC GARDEN PLANTS

Azalea	Ivy
Box	Jasmine
Clematis	Lily
Cyclamen	Lupins
Delphinium	Peony
Foxgloves	Yew
Hellebores	Wisteria
Holly	

For a comprehensive list visit
www.fabcats.org/owners/poisons/plants.html

Caring for your cat

Establishing daily care routines and regular health check-ups at the vet's will help to ensure your cat's well-being. Learning to recognize signs of ill-health or parasite infestation and how to deal with them are essential, and by grooming your cat regularly you will keep his coat glistening and your relationship strengthened.

Choosing a vet

Most cats find travelling and strange places stressful. Their misery is usually intensified in a clinic where they meet new people and other, unfamiliar pets. However, veterinary health care is essential, so making the right choice of vet will benefit everyone.

'Shopping around' – strange people, smells and sounds all combine to make a trip to the vet's especially traumatic for both kittens and adult cats. Therefore, it is important to research your area and choose a really 'cat-friendly' veterinarian. In addition to asking cat-owning friends and neighbours for recommendations, a number of things can help your assessment:

- Check websites for advice as well as cat-friendly commendations
- The waiting room should boast a good range of literature, such as information leaflets and magazines devoted to cats, with advertising posters highlighting medications especially for them. Feline-specific diets and equipment, such as tablet administering aids and dental care products, should be available
- Avoid waiting rooms without separate areas for dogs and cats, unless space is very limited
- Shelves for cat baskets or extra seating so carriers can be placed in elevated locations not on the floor are a must – remember, when anxious or scared, cats need to get up high
- Staff should show interest in both you and your pet, taking as many details as they can about his medical history and past experiences as possible, even if he's still a kitten
- Ask about open days and information sessions, because more clinics are organizing kitten introductory evenings and providing opportunities for owners to look behind the scenes
- Also check out emergency cover, asking for telephone numbers, addresses of other service providers and maps so that lack of information won't delay you getting help for your cat if you have a problem out of hours. Then keep them in a safe but easily accessible place so they are handy if this happens!

Preparation for the 'vet's'

Whether a clinic trip is routine or urgent, cats usually fare badly
if their owners don't handle them well. Many pets become
increasingly difficult with each distressing experience. There are
a few ways to help reduce your cat's stress levels.

CAT CARRIERS CAN BE 'FUN'

Open or remove the door, or the top half
if you can, and place his favourite blanket
inside to reduce unfamiliarity

Place the basket in a location your cat likes
to encourage him to use it

Hide treats and/or interesting toys inside

Feed him titbits, then his meals in its
vicinity, then inside, or play with him
nearby to create positive associations

Move it around the house repeating the
pleasant experiences in different areas

When he is happily accustomed to it, try
reassembling the carrier and shutting your
cat inside briefly, again making this as
positive as possible

Get him used to being moved around and
let out in a casual way

Positive associations – cat baskets are
generally stored out of the way between
these special, invariably negative, trips.
Keeping them as part of the furniture and
creating good associations can make
journeys, whatever their destination, less
stressful (see box, left).

Preparing to leave – cats really do seem
to have a sixth sense when it is time to
leave for their appointment: they will have
mysteriously disappeared from their usual
haunts and may be found in especially
inaccessible places. To avoid the stressful
handling that inevitably ensues, try the
following tips well before you are due
to leave:

• Get or keep him indoors in good time
• Stay calm or he will sense your anxiety
• Keep stressful handling to a minimum
• Tape down his blanket in his carrier so that
  it doesn't slip about

- Cover the carrier with a cloth to reduce the stressful impact

Stress relief – you can make his travelling kit smell reassuring with a commercial pheromone spray (see pages 66–67) or by rubbing the inside with a cloth onto which you've gathered your cat's own scents. Also spray a covering cloth to surround him with a stress-reducing atmosphere.

First impressions

You've made it to the vet's! To ensure that you get the most from your visit, it helps to know what to expect and to be prepared with any questions and to take notes if required.

KITTEN CARE CHECKLIST

Vaccinations What shots, when and how often are boosters needed?

Worming Type of medication (tablets, paste or liquid or injections), how frequently should your cat be wormed?

External parasite control What applications are needed, at what intervals and is environmental attention needed at home?

Diet What diet is recommended and how often should the kitten be fed now and in succeeding weeks/months?

Dental care What equipment, for example special toothpaste and a handled brush or a special ribbed, rubber cap that you place over your index finger

Identification Ask about microchipping as an additional means of identification to an ID tag on a collar

List your cat's symptoms – plus any questions you want answered, taking your pen and paper to note what you're told. Forgetting is easy in the heat of the moment! Also, file your cat's documentation and details of his medical history to take with you to each visit.

The full physical – when getting to know your cat, a veterinarian will want to make sure he's healthy and gather baseline data for future reference. Expect the clinician to examine your cat by looking into his mouth, eyes and ears, checking his skin and fur, feeling (palpating) his abdomen but listening (auscultation) to his heart and respiratory system. Cats are routinely weighed and some clinics have equipment to check blood pressure.

Questions, questions – this is your chance to ask advice about anything related to your pet's management and care. Don't be shy; also request demonstrations to start you off. See box left for what questions to ask when you take your kitten or your new adult cat for his first check-up.

Specialist clinics – if your cat is older, enquire about specialist clinics, particularly if something has been identified with which you may need ongoing help and support. Dental, diet and obesity clinics and senior pet sessions are all commonly offered by veterinarians now.

MICROCHIPPING

A microchip implant with a unique identification number is registered to your pet and your details are added to a national database. The tiny microchip is injected under the skin in the neck area and is similar to having a vaccination. If your cat is found, a scanner is used to read the chip and trace the owner.

Vaccinations

Fortunately, there is a range of effective vaccines against some of the major infectious diseases from which cats suffer. Vaccines licensed for use vary between countries and certain diseases are a risk in some areas of the world but not others, for example the UK and Australia are currently rabies-free. Regional and lifestyle (indoor-only versus indoor/outdoor) differences also influence the vaccines that are needed by pets. Consult your vet, who will be aware not only of your cat's individual circumstances but the conditions prevailing in your locality.

Feline panleucopenia (feline infectious enteritis or FIE) or Feline parvovirus – the frequently fatal severe form of gastroenteritis caused by this virus usually causes rapid deterioration and sometimes sudden death. High-level protection is afforded by vaccination.

Feline herpesvirus (FHV-1) and feline calicivirus (FCV) – these viruses cause most cases of cat flu, symptoms include:

- Conjunctivitis (inflammation of the eyelids)
- Sneezing
- Runny eyes and nose
- Raised temperature
- Sore throat
- Coughing

Infection, which varies from mild to severe, can result in long-term side effects and is occasionally fatal in vulnerable cats, especially elderly pets and kittens. Affected individuals can also become healthy carriers so that they pose a risk to other cats by shedding the virus intermittently, for instance if stressed. Whilst it is not 100 per cent effective, vaccination is generally effective in reducing the incidence and severity of cat flu infection.

Feline leukaemia virus (FeLV) – direct contact between cats is required for this virus to be contracted. However, once acquired it causes lifelong infection with most sufferers dying from tumours, progressive anaemia or immune compromise within three years of

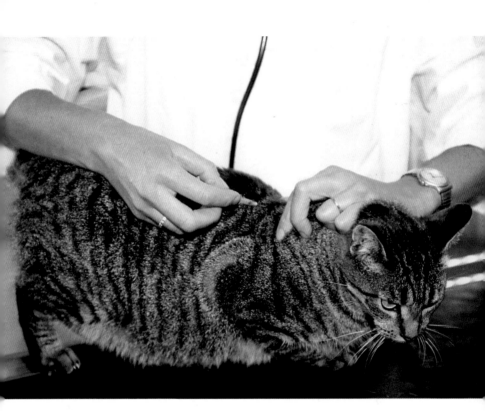

diagnosis. Although after vaccination not all cats will develop immunity against the disease, it is advisable to vaccinate any pet that is at risk.

Feline chlamydophilosis – protection against infection with *Chlamydophila felis* is not guaranteed after vaccination but the conjunctivitis, nasal and ocular discharges

and sneezing caused by this specialized bacterium may be milder as a result. The disease is most commonly seen in multi-cat homes with young cats and kittens generally being most vulnerable to infection.

Feline bordetellosis – a bacterium, *Bordetella bronchiseptica*, is one of the main organisms responsible for kennel cough in dogs. It can also affect cats causing the unpleasant signs, such as nasal discharges and sneezing, associated with upper respiratory tract infection. It can spread between cats and dogs and is most commonly seen in multi-cat homes. Generally infections respond well to supportive treatment, including antibiotics, although sometimes more serious pneumonia is seen. An effective nose drop vaccine is not routinely given but may be suggested by veterinarians in certain high risk situations.

KEY QUESTION

What happens if I miss my cat's booster vaccination?

Your cat will be vulnerable to the diseases against which the vaccination should protect him if he comes into contact with an infected pet or contaminated environment. If you delay too long your vet will probably advise that you start all over again with an initial vaccination course to ensure the best protection for your pet.

ADVERSE EFFECTS

Some people avoid vaccinating their cats because of concerns about post-vaccination reactions. Whilst adverse reactions can sometimes be serious, and not all vaccinated cats develop the anticipated immunity, expert advice supports the value of routine vaccination. It is, however, recommended that each cat's circumstances be taken into consideration when initial vaccinations are given and boosters to maintain immunity are discussed with the veterinarian at your pet's annual health check.

Rabies – the virus that causes this incurable disease is transmissible to almost all warm-blooded species, including cats and humans. Transmitted via infected saliva, it usually results in behavioural changes because it affects the nervous system. Cats travelling abroad can now be vaccinated against rabies.

Feline immunodeficiency virus (FIV) – similar to HIV, this purely feline infection is generally spread through cat bites causing a range of symptoms, including lethargy, weight loss, diarrhoea and chronic infections, for example of the gums.

Feline infectious peritonitis (FIP) – although the coronavirus (FCoV) that causes

this complicated disease is widespread, cases of fatal infection are uncommon. Symptoms include mild upper respiratory infection: sneezing, watery eyes and nasal discharge, and some cats may have diarrhoea, weight loss and lethargy. Households where multi-cat groups exist are most at risk. Vaccinations aren't available in every country.

External parasites

As a number of parasites affect cats, daily checks and routine home-based control regimes are essential. Your veterinarian will advise on local problems and supply the most effective products.

Treatment options – There is now a wide variety of different products and application options available, ranging from spot-on or drop preparations, specialized collars, powders, sprays and shampoos. Your veterinarian will help you to choose the best option for you and your cat.

Ringworm – This is actually a misnamed fungal disease that appears as circular patches of hair loss with crusting, which may not cause itching. It contaminates the environment and is highly transmissible to people. Immediate veterinary advice is essential and a topical cream treatment is often prescribed.

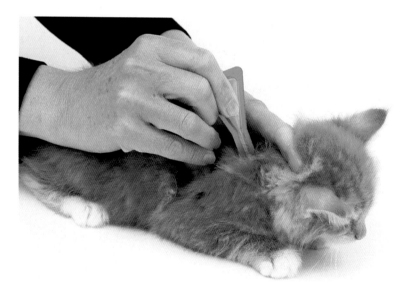

EXTERNAL PARASITES

Parasite	Control
Fleas are commonplace. Their numbers increase in hot weather and when indoor temperatures are high. Fleas are tiny, dark brown insects that hop or scurry through a cat's fur but their droppings (black grit) often indicate infestation. Some cats become allergic to flea saliva and the associated self-trauma can be intense	Cats catch fleas directly from infected animals. However, as larvae hatch from eggs deposited in the home, environmental contamination is also an important issue. Use a preparation that controls this in addition to keeping your pet flea-free
Ticks from rough grass attach firmly to a cat's skin during their feeding season. Starting small, they swell to a typical sack like-appearance of variable colour (slate grey to light brown)	Consult your clinic or attempt removal, if you know exactly what to do, as failure to remove all the head can lead to infection. Apply surgical spirit or alcohol, grasp and twist firmly with your fingers or use a special tick-removing hook available from vets. Tweezers increase the risk of leaving fragments behind
Mites live on the skin's surface or in the hair follicles; **ear mites** cause irritation and crusty, dark wax; **Cheyletiella mites** may not cause cats to itch but are identified by their characteristic dandruff; **harvest mites** are tiny orange mites that cause irritation between the toes and in the tiny pocket in the earflap. They are only seen in particular locations in late summer and early autumn	Veterinarians use an auroscope to see ear mites but diagnose and identify other mites microscopically using skin scrapings. This helps determine the best treatment regime, which must be thorough and may need to be repeated to resolve the problem. Where possible prevention is always better than cure

Internal parasites

Many cats do not show any signs with internal parasites, although some become ill or develop diarrhoea. Ask your vet or veterinary clinic staff about routine prevention or for specific treatment if infestation occurs.

Roundworms – white, round and up to 15 cm (6 in) long, pregnant and nursing queens pass these common worms to their kittens. Heavy infestations cause a pot-bellied appearance, debility and diarrhoea. Some affected cats vomit or pass worms in their stools. All pregnant females and kittens from 14 days old should be wormed.

Tapeworms – flat, segmented worms are usually only seen as desiccated segments, which resemble grains of rice, where a cat has been sitting or around his bottom. Fleas are the intermediate host for the most commonly encountered feline tapeworm, *Dipylidium*, hence the importance of regular flea and worm control. Another tapeworm, *Taenia* is contracted from infected rodents so, if your cat is a successful hunter, he may need more intense preventive treatment than less talented pets.

Lungworms – can cause respiratory symptoms, such as coughing, and are identified by special tests and require specific treatment.

Whipworms and hookworms – these intestinal parasites are not a problem in many parts of Europe but are found for instance in some places in Australia and the USA. Veterinarians can advise about prevention in affected areas.

Heartworms – cats are more resistant to infection than dogs and as this parasitic worm requires a mosquito to complete its

TOXOPLASMA

Many cats become immune to this intracellular parasite and show no signs. Scrupulous hygiene, particularly when disposing of feline faeces and litter tray contents, is essential. Pregnant women and immune-compromised people should take special care.

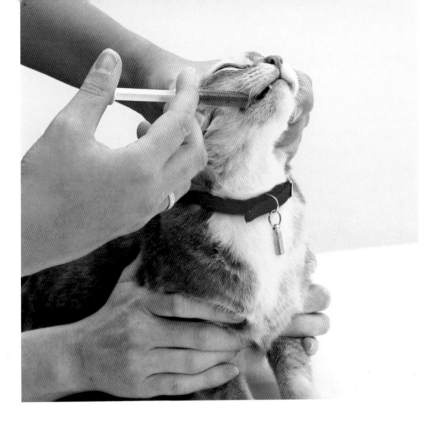

lifecycle the problem is generally restricted to outdoor pets in areas where infection is endemic, in the USA and Canada for instance.

Giardia – a single-celled parasite, which causes diarrhoea, is becoming increasingly common particularly when cats live in poor conditions. Veterinarians use faecal tests to diagnose it and treatment is available.

TREATMENT OPTIONS

Tablets
Liquids
Pastes
Spot-on preparations
Injection (tapeworm only)

Neutering

Cats become sexually active after puberty, which they reach around six months of age, although some individuals and pedigree cats may mature earlier. Whether you have a male or female cat, you need to decide whether to have them neutered.

Timing – because of variable maturity, females tend to be neutered from five months of age and males around six. In order to control the very high feline population density, rescue societies are, however, increasingly tending to opt for de-sexing both at an earlier age.

Castration – under anaesthetic the testes are removed from the male through an incision in the scrotum. This does not need to be sutured.

Spaying – ovario-hysterectomy in females, which involves removal of both ovaries and uterus, through a flank or mid-line abdominal incision that is then closed with stitches, is called spaying. Some veterinarians use dissolvable sutures. Others prefer the non-absorbable type, which means a visit to the clinic will be needed for them to be removed and your pet checked over post-operatively.

Aftercare – it is important that cats do not chew their operation sites, which can make the skin sore and introduce infection. Females sometimes remove their own

---

REASONS FOR NEUTERING

Population control

Avoid the vocalization, 'caterwauling', associated with feline mating seasons

Reduce the tendency to roam and fight for territory, mates and other resources

Reduce the risks of transmissible diseases, such as FIV, which has a higher incidence in free-living tom cats

Eliminate the characteristic very strong and pungent odour of mature tom cat urine

Avoid the risks associated with pregnancy and birth

Neutering

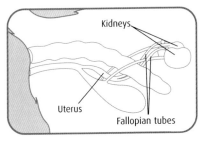

Queen

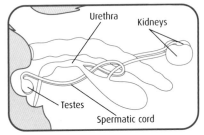

Tom cat

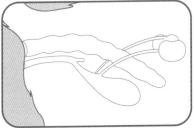

Spayed female

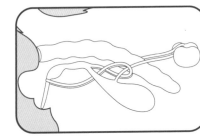

Neutered male

sutures too early and occasionally their wounds require resuturing. To avoid problems, vets often fit an Elizabethan collar (a plastic cone worn around the head), to prevent the wearer reaching the wound. When wearing one, cats must be kept inside.

After effects – de-sexed cats of both sexes tend to put on weight more readily than their entire counterparts so it is important to control your neutered cat's food intake.

KEY QUESTION

Should I let my female cat have a litter before being spayed?

There is no demonstrable value to this and even today pregnancy and birth are not without their risks to the mother. Sadly, there are already too many lovely cats seeking too few good homes so there is no reason for your cat to have kittens unless you really want her to.

Breeding

For entire cats, the urge to procreate is strong and it is possible for a female to have between two and three litters a year, with between one to eight kittens per litter. Being sexually mature particularly affects tom cat behaviour.

In season – female cats are seasonally polyoestrous. This means there are several oestrous periods each breeding season, which is brought on by increasing day length. The male will only be accepted during three- to four-day oestrous periods – 'being on heat' or 'in season' – characterized by vocalization, crouching with head down and paddling hind legs extended. At other times, even though behaviour may change, with more restlessness, rubbing, purring and stretching, advances are rebuffed.

Sexual maturity – un-neutered males develop pronounced secondary sexual characteristics becoming 'muscled up', with fat cheeks and pungent urine. They stray further and frequently away from home, mapping out territory in readiness for the mating season.

Mating – once the female signals she is ready to accept him, the male will mount from behind and grab her scruff to stop her attacking him. This and the barbs on his penis stimulate egg release – cats are induced ovulators. Mating can then be repeated several times over three to four days so, unless circumstances are controlled, a litter of kittens may have more than one father. Pregnancy begins with fertilized eggs implanting in the uterine horns.

Pregnancy – during the gestation period of around 63 days it is important that a pregnant queen gets good medical attention and a specially balanced diet to ensure she stays healthy and her kittens develop normally. Some learning takes place *in utero*, so getting the right nutrients helps set both their mental and physical development off to a good start.

Birth – as birth approaches the queen seeks out a good nesting site, which isn't always the place owners have dedicated to her! Here she will prepare a cosy nest, which signals that birth is imminent. Kindling, as it is called, is generally trouble free but if you have any concerns about your cat immediately contact your veterinarian. Once the kittens have arrived their mother will settle down to feed them, letting down her milk in response to their purring and kneading her abdomen. Seek attention from your vet if your cat or kittens are distressed.

Good health check

Check your cat daily by looking him over. Gently lift his tail to examine his bottom and run your hands carefully over his body. Don't forget his underside. Make sure he has no swellings, lumps, scabs, wounds, signs of bleeding or pain.

SIGNS OF GOOD HEALTH

Signs of good health	Cause for concern
Open, bright eyes	Dull, red, sore or discharging eyes
Clean, moist nose	Dry, crusty nose or any sort of nasal discharge
Glossy, smooth coat	Dull fur, dandruff or bald patches, red, scabby or crusty skin
Normal grooming pattern that keeps his fur looking good	Itchy, scratching and grooming frequently
Normal appetite	Off his food or eating more voraciously
Normal thirst or any increase can be explained by hot weather or high household temperature	Drinking more or has changed his fluid intake pattern, say from milk to water
Stable weight, normal body shape	Loses weight, shows too great an increase or develops an unusual shape, so that, for instance, his abdomen looks pear shaped
Alert attitude, interest in the environment, normal activity pattern	Becomes lethargic, hides when there is nothing to worry him, avoids contact with familiar people and pets he usually likes. Winces, hisses or spits when you pick him up (unless he normally resents handling)
No stiffness, for example when getting up	Struggles to rise or reach his usual high perches
Normal ability to run, play and jump	Limps or cries out when walking, or is disorientated or confused, collapses or loses his balance
Clean face and chin	Dribbles, claws his face or eats with difficulty or on one side
Normal respiratory rate, good recovery from exercise, plenty of energy	Coughs, sneezes, wheezes, has laboured breathing
Passes normal faeces and has no signs of distress on urination	Is constipated, has diarrhoea or difficulty passing urine (emergency in males)

Medicating made easy

If your cat requires medication you need to be able to administer it correctly. The secret of efficient treatment is knowing 'how to' and being well organized. Don't be afraid to ask staff at your veterinary clinic for demonstrations before you attempt it.

Dosing your cat – there is a range of feline-friendly medications available now and more products that can be given in a favourite food. Discuss the options with your vet.

WITH AN ASSISTANT

1 Restraint – this is the key to safely administering medication, such as pills, liquid formulas and eye or ear drops, if you have to. Place a thick towel on a table and put your cat in the centre, head towards you.

2 From behind, the assistant should lean over the cat, squeezing his behind gently between their elbows and firmly grasping the cat's front legs.

3 If you are right-handed, place your left hand over you cat's head and gently tilt his head backwards. Pop the tablet into his mouth before closing it and rubbing his throat to encourage swallowing.

SINGLE-HANDED

1 Spread a towel out and place your cat in the middle. Fold one corner crossways over his opposite shoulder to enclose his front leg. Repeat with the other corner and leg.

2 Fold the towel over his back in a similar fashion so he is quite tightly restrained.

3 Follow Step 3 left to administer the pill. Alternatively, sit down and with your cat's bottom towards your body put him on your lap and lean over to reach his front end.

Eye drops – if he isn't cooperative follow the steps to restrain your cat (see opposite), then rest your hand on the top of your cat's head, which helps avoid contact with his eye because you can feel if he moves. Squeeze the medication onto his eyeball and gently massage the eyelids together.

Ear drops – restrain your cat following the steps opposite, then grasp the affected ear in one hand and gently place the ear drop bottle into the ear canal and squeeze out the requisite dose. Massage the cartilage cone at the base of the ear to ensure the liquid reaches all the areas of the outer ear.

WARNING!

Never give your cat human medicines without consulting your veterinarian because common drugs, such as paracetamol, are toxic and can be fatal.

Accidents and emergency

Organize an emergency kit and immediately replace anything that's used. Include cash for a taxi, even if you normally drive. Misfortune often strikes when cars are inconveniently unavailable.

YOUR CAT'S FIRST AID KIT

Cotton wool – to staunch bleeding with a pressure pad and bathe cuts

Sterile dressings

Several bandages of different widths

Blunt-ended scissors and tweezers

Adhesive tape to secure dressings

Antiseptic wipes or solution (table salt can provide a dilute saline solution – use a pinch in a cup of warm water to wash minor wounds)

Two pairs of disposable gloves

Torch

Strong, medium-sized blanket – used to safely handle a distressed cat that's uncharacteristically aggressive, for moving an injured pet and to keep him warm

Action stations – if your cat is injured stay calm. When shocked, scared and in pain, even gentle cats can become self-protective. Keep handling to a minimum, unless the situation is life threatening or your pet is in a dangerous position. Then move him gently to a safe area, if necessary by rolling him carefully onto a blanket or by using it to protect yourself. Lay him on his side and remember the importance of ABC:

- **Airway** – using latex gloves scoop away excess saliva with your finger, if his airway is obstructed. (Do this only if necessary and safe. For instance, if your cat is struggling to breathe because his mouth is full of discharges or he is unconscious.)

- **Breathing** – check his chest is moving normally. Try artificial respiration only if he has stopped breathing. Close his mouth and with your hand cupped around the top of his nose breathe into his nostrils. Pause and repeat until he starts to breathe again.

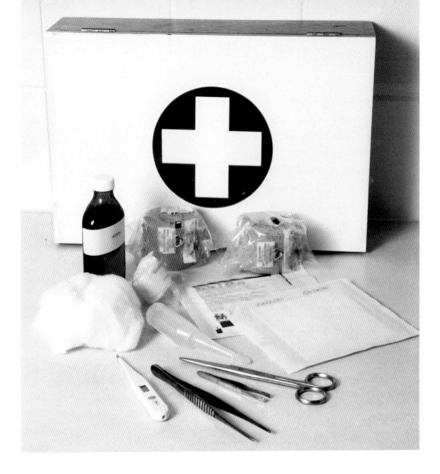

- **Circulation** – if a cat's heart has stopped, massaging his chest may help start it again. Place one hand on his back and the other around his lower ribcage with your thumb on one side, fingers on the other. Firmly pump (don't crush) two or three times, then administer artificial respiration before pumping again. Do not panic (no matter how distressing the circumstances are). Keeping a cool head and gently working through the necessary endeavours is essential.

Living dangerously

Some cats use up their nine lives; others lead mishap-free existences. Whilst hoping your pet falls into the latter category, it's best to be prepared just in case. Following an accident, telephone your veterinarian for advice as soon as your cat is in a safe position and you have administered any necessary first aid. Do not offer food or drink, as your pet may need an anaesthetic.

Accidents – after a road accident or fall apply the ABC principles (see pages 116–117), then assess the damage. Your cat may be shocked. Don't be surprised if he shivers, has cold extremities and his gums are unusually pale. If he has any wounds or broken bones, handle him with extreme care. Gently place him in his carrier in a darkened room, cover him with a blanket to keep him warm and contact your vet.

Cat fights – never be complacent if your cat has a fight as minor puncture wounds can become infected very quickly. When he has calmed down, provided he is cooperative and you can safely do so, clip away the hair. Bathe the wound with warm water or dilute saline and dry the area well. Seek advice.

Bleeding – treat minor wounds as you would a cat-fight injury. If your cat's wound is bleeding heavily, apply pressure using a cotton wool pad well covered with gauze to avoid leaving fibres in the wound. Press down on the affected area, confine your cat to his carrier and take him to your clinic.

Burns – don't apply cream to burns or scalds. Try to cool the area by covering with a wet cloth or an ice pack (a bag of frozen peas works), well covered to prevent chilling the tissues excessively. These injuries may look minor but always seek a professional opinion.

Electrical accidents – do not touch your cat if he's unfortunate enough to be electrocuted. Switch the source off at the mains or use a wooden stick or broom handle to push him away from it. Apply ABC principles if necessary and seek veterinary help immediately.

CAT BITES

Even healthy cats carry a range of bacteria in their mouths, which although not generally harmful, can cause severe infections in other individuals, including people. Any wound inflicted by a cat should be treated carefully and medical advice sought if it is serious or someone is particularly susceptible because, for example, they are immuno-compromised.

Grooming

Cats are proficient groomers. The rasping action of the cat's barbed tongue removes loose hair and stimulates the skin's natural oil production, to help with waterproofing and insulation. Sometimes, however, cats need extra help with their coat care.

Regular grooming – keeps your cat's coat in good condition, but has other benefits, too, from providing mutual bonding time with your pet to giving you the opportunity to check him over for anything abnormal, such as wounds, lumps and parasites (see pages 104–105). Ill or elderly cats will also benefit from a little assistance.

Coat type – longhaired cats require daily grooming sessions to prevent knots and tangles forming, especially behind their ears, inside the groin and around the rear end (see below). Groom shorthaired types weekly. Start by brushing from your cat's head to his tail working with the fur to remove dead hair. Then use a comb as illustrated above.

LONGHAIRED CATS TECHNIQUE

Concentrate on grooming with a wider toothed comb and slicker brush, gently teasing matted fur apart with your fingers. Only attempt to cut out tangles if your cat is very cooperative and will stay still. A useful tip is to bury your comb in the base of the mat to protect the skin then you can snip away safely. Comb out the barbered remaining fur to prevent the mat growing back again.

Fur balls – especially when moulting (see right), cats tend to swallow hair when grooming, which can accumulate in their stomach and intestines forming fur balls. These can cause vomiting, constipation and very occasionally intestinal obstruction. Regular grooming helps to reduce the incidence of fur balls.

Early start – starting kittens off early with little and frequent sessions makes lifelong grooming pleasant for both parties. Reward cooperation with prized food treats and end with praise and play.

Reluctant sitters – treat adult cats who have not been used to grooming in a similar way to kittens, but expect acceptance to take longer and keep initial sessions very short. Face your cat away from you if he is shy.

MOULTING

When the days lengthen in spring, cats moult heavily to get rid of the dead hair from their thick winter coats in preparation for warmer weather. However, because we tend to keep our homes much warmer when the temperature falls than people did in the past, many cats also shed significant quantities of hair in the autumn when we turn up our heating. So watch for excess moulting and be prepared to step up your grooming efforts at this time too.

Grooming kit – there is a huge variety of grooming tools available. Choose those that are suited to your cat's coat type, and include a brush and a comb. Some cats prefer grooming gloves or rubber brushes. You may also need nail clippers (see page 123).

Teeth and claw care

Tooth problems and gum disease not only are unpleasant for cats but can also lead to other infections. It is now recognized that everyday dental care is part of good management and there is a range of special toothbrushes and pastes available for cats. In addition, there are some specialist feline chews and diets, which may help to prevent the build-up of tartar on teeth.

Brushing your cat's teeth – little and often to get kittens immediately accustomed to the process is the best way forward. Start by gently examining your cat's mouth, lifting his lips and rubbing a finger along his gums and up and down, before you move on to using whichever toothbrush seems best in your situation. This may take a few days with an adult cat unaccustomed to tooth-care regimes but choosing a good time when your pet is calm should help. Choose a paste with a flavour your cat will like and begin with his easily accessible front teeth. As he learns to accept the procedure move further back until you can remove plaque and debris from his whole mouth. If you are successful you should be able to reduce the risk of your cat developing gum disease and the amount of veterinary attention with scaling and extractions under general anaesthetic that he needs during his lifetime.

CLAW CLIPPING

You should check your pet's claws regularly as part of his general welfare routine. Not all cats need their claws cutting but indoor-only and senior pets may get caught by their long nails in bedding and carpets if they fail to wear them down through routine use.

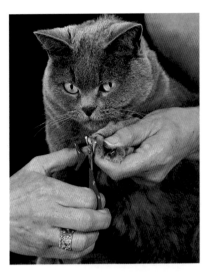

1 Start by restraining your cat (see page 114) or sitting him comfortably on your lap.

2 Grasp one paw and gently extrude a nail by exerting pressure at the base.

3 Use sharp clippers to snip off the white tip about halfway up. You must avoid cutting the quick, a triangle of pink sensitive tissue with blood vessels at the claw base. You may prefer special clippers that encircle the cat's claw. When you've finished reward your cat.

Bathing

The fine balance of oils in a cat's skin and fur and their expertise at self-grooming precludes the need for regular bathing. Bathe your cat only when strictly necessary, for example if your cat has been in contact with something like oil that has contaminated his fur or he has a skin problem that requires shampooing with a medicated preparation.

Why don't cats like water?

Originating in the harsh, dry African savannah, cats obtained enough fluid from their food and didn't need to drink as well. Water is often quite dangerous to be around and sadly, although most animals can swim if they have to, cats do sometimes drown if they fall into deep water. It's hardly surprising, therefore, that many cats avoid water. However, some bold individuals do enjoy playing with it and the Turkish Van breed is sometimes referred to as the swimming cat because of this ability and its liking for water.

Professional advice – unless your cat has soiled himself, perhaps when travelling, you should always consult your veterinarian rather than tackling the situation yourself. Some toxic substances can be absorbed through the skin and may need specialist attention at the clinic.

Negative associations – most cats do not like to be restrained and fewer still enjoy being bathed. If bathing proves really unpopular with your cat it may be better for your relationship if he associates it with the professionals rather than you.

BATHING TIPS

If your vet has advised you to clean the contaminant off your cat's fur before rushing him to the clinic, talk calmly when bathing for reassurance and remember the ultimate customary reward

Ensure the bathroom or kitchen is warm and your cat cannot escape

Have plenty of warm water, towels and your special cat shampoo readily to hand

Place a non-slip mat in your bowl or basin for your cat to stand on

Use a sponge or showerhead for rinsing, but test the temperature first

Wet your cat's coat thoroughly taking care not to get water in his eyes or ears

Apply a little shampoo and massage it carefully into the fur

Rinse well with plenty of clean water

Repeat as directed

Squeeze away as much water as you can before towelling your cat dry

Keep him warm and don't let your cat out until he is quite dry

Fitness and happiness

In order for your cat to feel settled and content in his home and to live a stimulating and happy life, you need to fulfil his basic daily requirements. An appropriate diet with careful calorie control is as important as spending time playing and interacting with your cat.

General well-being

To keep your cat in peak condition you need to consider all aspects of his well-being – both physical and mental – and learn to recognize the signs of anxiety or unhappiness. Being aware of his temperament, age and environment will all help.

Active lifestyle – keeping your cat in good bodily condition, aided by a healthy diet and plenty of opportunity for exercise, will ensure his general physical well-being. See pages 130–133 for advice on diets and general feeding regimes.

Emotional well-being – cats are sensitive creatures and it is not only their physical health but also their emotional happiness that we need to consider. Being aware of what causes anxiety, their needs and temperaments helps owners ensure their pets have happy and active minds. (See also pages 144–145.)

Life stages – at different stages in his life, your cat will need different physical and mental stimulation. If for any reason your cat was deprived of socialization at an early stage, for example, it will affect his outlook on many aspects of his home life.

Suffering in silence – the natural feline way of responding to difficulties is to retreat and hide, so many distressed cats are never brought to the attention of veterinarians and behaviourists who could help them, because their behaviour goes unnoticed. Look for signs and seek professional help if required (see page 168).

One day old   Ten days old   Three weeks old   Five weeks old   Eight weeks old

High risk conditions – more cats than ever are living in high risk conditions, perhaps where too many cats live in too small a space for comfort, they are restricted indoors throughout their lives or they are part of a feline household that is made up of unrelated adults cats. Be aware of these outside factors on your cat's mental state.

FELINE LIFE STAGES

Neo-natal period	0–10 days
Transitional period	10–21 days
Socialization period	2–7 weeks
Juvenile – end of socialization period	Sexual maturity (from 18–20 weeks)
Adult	from 10–12 months
Mature	7–10 years
Senior	from 11 years

Fourteen weeks old   Five months old   Adult cat

Feline food facts

Everything – physical condition, mental development and general good health – depends upon well-balanced nutrition appropriate to a cat's species and age. Throughout cats' lives their nutritional needs change and your veterinarian is there to help you.

What cats need – their species is developed to live on a diet primarily composed of small rodents. Therefore, cats gain their energy from protein and the feline requirement for carbohydrate is low. But they are obligate carnivores, which means that they must eat meat because of the essential nutrients it contains that they cannot get from other

sources (see box). For example, cats cannot manufacture taurine, a crucially important amino acid without which they develop blindness and heart problems.

Balanced diet – your pet's calorie requirement will vary according to his age, health status and lifestyle. So while it may be tempting to prepare home-cooked food for your cat on a regular basis, providing the right quantity of a balanced diet is a challenge and the consequences of getting things wrong can be serious.

Prepared cat food – this is the ideal way for most owners to feed their cats, either in a wet (tinned or packaged) or dry (biscuit or kibble) form. These products are specially developed to provide the ideal range of nutrients for various feline life stages and they come with calorie guides that indicate the correct amounts for individual pets.

Special diets – in addition, a whole range of prescription diets is available to help support the medical care necessary to keep cats suffering with such conditions as diabetes, kidney failure, heart problems or dietary sensitivities, in optimum condition. Special convalescent diets aid recovery after cats have had debilitating illnesses or accidents.

ESSENTIAL NUTRIENTS

Cats must obtain the correct amounts of the following nutrients from their diet to maintain optimum health:

High levels of protein – although a cat's age and health status may dictate that this is lowered. Always ask your vet to advise you

Taurine (an amino acid) – for general bodily functions, especially for the heart and eye health

Arginine (an amino acid) – to regulate ammonia levels

Arachidonic acid, linoleic acid and other essential fatty acids – for general good health and to maintain a range of important bodily functions

Vitamin A – which cats are unable to manufacture themselves, for healthy skin and eyesight

B group vitamins – for general health

Fibre – which cats would get naturally from consuming the whole of their prey, helps to prevent constipation by maintaining normal intestinal activity

Feeding regime

Your vet will be able to advise you of the best type of diet suited to your cat, depending on his age, breed and general health. Once you have established a regular feeding regime and a brand that your cat likes, try not to introduce any unnecessary changes.

FEEDING YOUR CAT – DO'S AND DON'TS

Do	Don't
Ask your veterinarian's advice or for a referral to a qualified feline nutritionist	Feed any kind of bones – they soften and can shatter when ingested, sometimes with severe medical consequences. The risk is highest with poultry
Feed a diet specifically formulated for your cat's age and health status	
Make sure you know his weight and stick to the quantities advised by carefully weighing or measuring each day's portion – estimates invariably lead to overfeeding	Feed raw chicken wings or necks to aid dental care – this can be risky with cats that cannot chew well, for instance. There is also the potential problem of bacterial contamination
Use a reputable source for any home-prepared meals you give your cat	Give milk to adult cats. They lack the enzyme to digest lactose (milk sugar) and it often upsets their digestion. If your cat enjoys milk, use only a specially formulated 'cat milk' that is lactose-free
Cook food well to avoid bacterial infections – food poisoning is a risk with raw food diets, even when human standard ingredients are used	

Dry diets – have the advantage they can be left down for cats to snack. For many active cats this is ideal, although it is advisable to put out smallish quantities, as the oils dry foods contain evaporate, reducing palatability.

Wet food – can be fed several times a day, although two to three is the general rule – but don't exceed his daily calorie allowance and always ensure it is at room temperature not straight from the fridge. Cats like their food warm (they developed to live on a diet of freshly killed prey), and its palatability is enhanced by gentle warming. Check it is evenly heated and not too hot. Chilled food can sometimes cause intestinal upsets.

KEY QUESTION

What are suitable treats?

Small pieces of chicken, tuna, fish and prawns are suitable treats and you can get dried versions of prawns and liver – only give small quantities. Avoid anything with sugar or that is very fatty.

Controlled amounts – if your cat is inactive or has a tendency to 'comfort eat' out of boredom or stress, carefully control the amount you feed each day taking care to follow the manufacturer's directions.

Avoiding obesity

Feline obesity is reaching epidemic proportions, bringing serious health risks, such as diabetes, heart disease, hypertension and osteoarthritis. Be strict with portion control and follow the tips below to keep extra calories under control.

Treats – there can be many factors involved in obesity but owners generally feature in its development. You may have a tendency to express your love through the provision of extra calories, which quickly accumulate as feline fat!

Automatic response – you may also misinterpret your cat's overture for attention or a game as a message that he is hungry. Obligingly, when you open the tin or food cupboard, he eats whatever you put down, which seems to confirm the misunderstanding.

Earning rewards – cats that fend for themselves make numerous unsuccessful hunting attempts before they get a meal. By making your pet work for his food (see box, right) and when you think he may be hungry, playing, petting or grooming to ensure you aren't just misinterpreting the situation, you will be on the road to success.

KEY FACT

Overweight cats are also at risk of developing feline lower urinary tract disease (FLUTD), a complicated syndrome that is often recurrent and undermines quality of life.

FORAGING GAMES

Hide-and-seek – dry food offers more opportunities but even with a wet-food diet, cats can seek out their treat. Split a modest quantity into minute portions and place them in several different locations

Fishing – place tiny pieces of chicken, tuna or prawn in greaseproof or rice paper parcels, attach to fishing rods and dangle away. Every now and then your cat must be allowed to pounce, 'catch' and devour the offering – it's a great way to bond as well

Trails – scatter dry food on a clean kitchen floor or patio; lay trails in corridors; place an individual piece on each step of the staircase

Pass the parcel – place a few pieces of fresh food in cardboard boxes with holes, wrapped in plain tissue in a cardboard tube or screwed up in brown paper bags

Chase – buy a commercial food ball or maze-style puzzle feeder, or make your own using a small, clean plastic drinks bottle, and fill with dry food that will fall out of the toy as it is chased around

Avoiding dehydration

Dehydration in cats often goes unnoticed, especially in dried food fans, older pets and house cats. If left untreated it can lead to serious health problems, such as feline lower urinary tract disease (see page 135). Cats can have quite particular drinking preferences – try to cater for your cat's whims to keep him in top condition.

TIPS FOR INCREASING FLUID INTAKE

Find out what your cat likes and increase his fluid intake by:

Collecting fresh rainwater, offering bottled or filtered water or running a tap

Offering a variety of containers

Placing containers in a variety of locations

Obtaining indoor or outside water features (so long as they are not contaminated with algae or chemicals) as these attract some pets

Purchasing a pet fountain – their noise helps blind and partially sighted cats find their water more easily

Adding more water to his wet food

Making meat or tuna juice ice cubes and placing them in dishes to melt

Variety – a bowl of fresh water beside his food dish can feel unnatural to your cat as his instinct is to drink from a variety of small pools and crevices where rainwater gathers, rather than waterholes where animals congregate. Give your cat several choices for his water supply in the home.

SIGNS OF DEHYDRATION

Many conditions can cause lethargy, confusion, dull coat, dry nose and sunken eyes. These can also be signs of dehydration and if you are concerned that your cat may be dehydrated you can try gently raising his skin over his back. If he is fit and healthy, it should immediately return to normal. If it stays up in a 'tent', taking a long time to slide back into position, promptly seek veterinary advice.

Location – experiment with different locations for water dishes, some elevated, and be aware of your cat's requests. A running tap in the bathroom or a small 'pool' in the sink, may be preferred over the static bowl in the kitchen.

Containers – some cats prefer wide shallow dishes where their whiskers do not touch the sides, others go for deep containers like tall drinking glasses, freshly filled flower vases or watering cans. Avoid plastic, which may gradually absorb stale, off-putting odours.

Rainwater – cats are often seen drinking from fresh rainwater puddles. If your cat prefers rainwater, collect some in clean containers to supplement his supply.

Encouraging activity

It is not just to promote physical fitness and avoid obesity that we should be encouraging our cats to keep active whatever age they are. As with people, it is now recognized that 'use it or lose it' mental activity and exercise are important in keeping your cat alert and young at heart.

Interaction – you can have great fun interacting with your cat when you try to make his lifestyle as healthy and interesting as possible. None of this needs to cost a lot – ingenuity and imagination are often the keys to success. You'll find plenty of interactive toys in pet stores or use whatever comes to hand, such as balls (some cats learn to retrieve balls if rewarded with food treats), toy fishing rods, feathers on sticks and twists of tissue paper that skitter across the floor.

Nurturing the mind – in addition to using routine activities, such as playing foraging games at mealtimes, to employ your cat's little grey cells help avoid monotony by:

- Rotating his toys so that they retain their novelty and his interest
- Opening a different cupboard each day to allow him exploration opportunities
- Hanging mobiles and glitter balls or old CDs for him to watch

WARNING!

Never play with fingers or toes – even with tiny kittens – as this will encourage your cat to view them as prey into which he will readily dig teeth and claws.

Play sessions – your cat may initiate games with you and you will soon learn when he is in the right mood! You can tempt your cat into a game but be prepared for it to be short, and on his terms!

Activity centres – you can purchase carefully selected activity items or ask someone who enjoys making things to construct climbing frames and activity centres to your cat's own specifications for maximum enjoyment.

EXERCISING THE BODY

Stable and safe things to run up, climb on, jump off or hide in are great for cats but so are:

Boxes

Tunnels made from fabric or cardboard

Newspaper or cardboard 'tents'

Paper carrier bags (remove handles for safety)

Baskets of different sizes and shapes, some turned over, some on their sides

Stable stepladders

Cleared shelf units

Training

Vets and behaviourists are fond of the phrase 'a cat is not a small dog'. You need to respect your cat for his feline ways, and be realistic in your expectations of how he will respond to training and commands. There are approaches and techniques that will help your cat to come when called and avoid potentially risky situations when necessary.

Alternatives – old-fashioned training methods based on punishment and fear have no place anywhere anymore. When it comes to training cats, particularly if you wish to discourage an action that is natural, try to provide an alternative or use physical barriers to discourage them. For example, if he jumps up onto raised surfaces like the kitchen units, provide plenty of elevated alternatives, or use double-sided sticky tape, which cats do not like walking on, along the edges.

Distraction – if your cat is heading towards danger or somewhere you'd rather he didn't explore, then distracting him is a great way of interrupting his journey, rather than becoming involved yourself by picking him up. Throw something onto the floor in the opposite direction, it will attract his attention and persuade him to jump down to investigate. This keeps you out of the

situation and preserves the remote 'Act of God' nature of the action. A sharp 'no' can also be a useful interruption.

Avoid contact – picking your cat up to remove him from a situation can give the wrong message (and you may be injured if he struggles). By talking to him ('don't do that kitty'), looking at him (eye contact is a reward) and handling him we are giving him positive attention. Try the methods above to avoid any confusion.

Clear signals – your cat will be confused if you're not consistent in your approach, for example getting cross when he pesters guests at the table but giving him titbits throughout your own informal meals. Don't change the rules and expect your pet to somehow know. Always use the same words for commands to avoid confusion.

This can be especially difficult if yours is a busy family home or you have lots of visitors. Make sure everyone knows your cat's house rules and agrees to abide by them.

Inconsistencies that arise when different people have different expectations of cats often cause feline stress and contribute to problem behaviour.

Coming when called – it is useful to teach a cat to come when he's wanted. Start at mealtimes to encourage compliance.

- Select your signal – the rattle of the dry food box or a fork tapped against the side of his tin
- Choose a verbal command – his name or 'here kitty', for example. Adopt a cheerful, light tone of voice
- Keep the command short

Do this every meal and it should soon be available for use at other times but always have a reward ready and don't misuse the command or its effectiveness will be undermined.

Clicker and target training – a clicker is a small plastic box with a metal insert that makes a sharp 'click' when depressed, and a food reward is offered. It is popular for training a whole range of behaviours. Some tricks help with handling pets; others are used simply to make life more interesting especially for cats kept indoors. Target training involves teaching a cat to initially touch a stick with his paw or nose, which then leads on to more sophisticated sequences of behaviour, such as climbing over agility courses made from furniture. With good technique, the right cat and lots of patience it can provide lots of fun. But never pressurize your cat to perform if it's not his thing.

TRAINING TIPS

You must always make it worth your pet's while to comply. Praise and encouragement are helpful but food rewards or play – whichever suits your cat – are essential.

Don't try to teach your cat a behaviour that contradicts his natural temperament. Make the most of his natural skills: carrying things in his mouth, batting with his paws or investigating with his nose.

THE THREE C'S

Successful training depends on:

Clear – be clear what you are trying to achieve and how to do it

Consistent – everyone must do the same thing all the time, especially important in a family home

Cat-orientated – cats must not be expected to do anything that runs counter to their species natural behaviour

Coping with change

Cats are creatures of habit and any changes to their daily routine or their familiar home territory can upset and distress them. They like to feel in control and, therefore, anything we can do to minimize the stressful impact of change is likely to be beneficial for their emotional well-being.

Everyday change – there are some everyday events that we take in our stride but their impact on our cats can be much greater. A change in your working hours, for example, may affect your daily routine: with later or earlier feeding times. A new cat food may

COPING WITH CHANGE CHECKLIST

It helps if we can anticipate changes and plan well ahead for them but sometimes life simply moves on. Whether it is an unexpected disruption or a long-awaited event, a number of things can help cats cope and settle quickly:

Keeping your cat's routines as nearly normal as possible

Making sure that his food, litter, treats and playtimes are all the same as he normally has and occur at roughly the same times of day – a cat that misses out on his regular playtime or little interactive rituals can be frustrated and upset

Maintaining your usual style of interaction – constant intrusive reassurance, while motivated by good intentions, often makes things worse

Ensuring that there are plenty of refuges and hiding places available and that no one invades a cat's space if he has retreated – you can always just sit quietly nearby and chat in a low-key, soothing manner to show him all is well

Avoiding removing items that have his own reassuring scents on them or those of someone he is especially well bonded to – scent signals are one of the most potent de-stressing mechanisms cats have

cause upset, or even the arrival of some new furniture. See the checklist below for how to help your cat at these times.

Familiarity – be sensitive about your cat's favourite sleeping places – make sure that he has access to at least two places. Try to avoid washing all his bedding at once, so that he has some familiar smells to reassure him.

Missing you – when cats bond closely with one of their carers and this special person is absent for a few days they will miss them. Help him cope by letting him have access to his special person's bed where their personal scent will be strongest, thus providing a calming oasis for a distressed pet. Or the 'favoured friend' can leave a worn garment, such as a T-shirt as a comforter.

Coping with stressful events

There are times when major changes occur in our lives or our homes that have a serious impact on our cats' lives, too. Anticipating potentially stressful events and taking advance measures is ideal in keeping your pet as happy as possible.

Changes in his environment – building work and redecoration bring with them strangers, noise and disruption, affecting access and escape routes and changing the scent profile and layout of your cat's familiar territory. Generally, a dedicated refuge area, as when he first arrived (see pages 70–71), is needed to protect him from the comings and goings.

New additions – remember also that exciting things can be stressful to pets, an expansion of your household because a baby is due or someone new is moving in, for example. Closing off areas of the home that will no longer be available to a cat before the new person arrives, prepares him beforehand so that both changes don't occur at the same time.

Multi-cat homes – the impact of everyday or more unusual events can be especially problematic in multi-cat groups, even when well-bonded cats live equably together. A particularly troublesome scenario is when one cat comes home from a visit to the vet's smelling of the clinic. Wiping the patient with a cloth before he reaches the clinic,

BEING PREPARED

Before a new person arrives, introduce their scent to your cat and associate it with something good – a special food treat or new toy – to help reduce any stress associated with new people

Consider playing recordings of babies' noises before they arrive if cats have never experienced living with them

Protecting pets from the unusual odours carried by the baby items you acquire in advance is wise, as these can become the target for marking behaviours in distressed cats

storing it in a sealed plastic bag to preserve his scent and then reanointing him before his return can be helpful.

New home – try to shield your cat from your packing by keeping doors closed and throwing a cloth with his familiar scent on it over boxes and cases, for example. If your cat is happy to go into a cattery, this is often best. Otherwise create a haven for him somewhere quiet with all the facilities he needs. Do the same in your new home, so that he can quietly adjust, and try to get familiar smelling items out for him before you let him out of his sanctuary to explore the house. (See also pages 80–81.)

New territory – following a house move, make sure that your cat is very well bonded to your new base before allowing him outside for short periods. Stay around to supervise and remember that he is now the 'new kid on the block'.

Holiday care

There may be times when you have to leave your cat for a few days, or weeks. Finding the right care for your cat, whether he will be cared for at home or in a cattery, is important for his health and happiness, and for your peace of mind.

HOLIDAY CHECKLIST

Leave full details of your cat's:
Veterinarian
Medical history and health problems
Dietary sensitivities
Routines (feeding times etc.)
Favourite foods
Handling preferences
Command words

Cat-sitters – for those pets that are timid, unsociable with other cats or territorial in nature, or cats that are very dependent upon human company, having someone familiar or very cat friendly to stay or visit in the home is generally the best option. The upheaval of leaving base and spending time in unfamiliar lodgings surrounded by strangers, two- and four-legged, can be emotionally traumatic.

Cattery – some cats are happy to visit a cattery where they will have human and feline company. Carefully check out the credentials of all concerned before you commit yourself. Catteries should be:

- Well run, by friendly staff who care for their feline charges
- Willing to allow a tour at a quiet time of day

COPING TIPS

Do not overwhelm your cat with extra cuddles before leaving and returning

Keep all routines as normal but minimize the impact of packing suitcases, for instance by shutting the spare room and keeping all the clutter in there

Do not wash any of your cat's equipment before you go. The familiar scents will be reassuring for him

Send him to the cattery with his own bed plus extra blankets so they can be washed in rotation

If someone is coming to care for your cat at home, introduce them ahead of time by asking for an old T-shirt that they have worn and associating it with your cat's favourite food

Leave (unwashed) items of your clothing in sealed plastic bags for 'topping up' your scent throughout your absence and leave your bedding where your scent is high

- Anxious to know everything about your cat that will help him settle
- Clearly balancing hygiene and disease control with a genuine regard for the emotional well-being of the cats

Problem
behaviour

Feline behaviour problems can strain
relationships between cats and
people, sometimes to breaking
point. You can avoid pitfalls by
being realistic in your expectations,
and getting to know your cat's
personality and background. The
following pages identify some of
the most common problem areas
and how to help you and your cat
deal with them.

Avoiding conflict

Doing all that you can to prevent or anticipate potential problem areas in day-to-day situations will benefit you and your cat. With a little understanding it is possible to avoid conflict and get help for your pet before the problem escalates.

HIGH RISK SITUATIONS

There are some circumstances where the risk of behaviour problems developing is increased. Identifying them and applying special measures can go a long way towards ensuring good welfare and reducing the risk of things going wrong. Potentially troublesome scenarios are:

Multi-cat homes

Indoor-only lifestyles

Cats living where the general feline population is high

Rescued, poorly socialized cats

Timid and fearful animals

Building/decorating work or moving house

Bereavement – loss of people and favoured pet companions

Consideration – being sensitive to what your cat requires is a great starting point. Before you bring a cat into your home make sure that they have all they need, relevant to their age and sensitivities.

- Enrich the environments both indoor and outside to meet all his needs (see pages 64–65)

- Have coping strategies for everyday stressors and more unusual events that may stress him (see pages 66–67)

- Make allowances for his natural behaviour (see pages 40–41)

Unwelcome behaviour – if your cat starts to behave in a way that is unacceptable to you, try to discover the stressors that may be triggering it, rather than getting angry or irritated. There may be a simple solution, or your cat may require professional help.

Medical causes – some diseases can profoundly change a cat's behaviour and in

other circumstances a medical and behavioural component combined together are responsible for the difficulties. Therefore, it is essential to always check out an affected pet's medical status when behaviour problems first emerge. Your veterinarian should also be able to offer 'first aid' behavioural advice.

Damage and destruction

Few things are more distressing than looking forward to having a
pet and finding you've got a home wrecker! A small cat can cause
a surprising amount of damage. Identify where his frustrations
lie and provide what he needs so that he can engage in all his
natural behaviours in an acceptable manner.

Exploring the world – cats are full of energy and need plenty of suitable outlets for it. They are bright and curious and have a tendency to knock precious items off shelves when they run, jump, climb and play. Although agile and light, they cannot be entrusted to always pick their way carefully through your china cabinet.

KEEPING OUT OF TROUBLE

Provide suitable scratching posts in areas where your cat needs to establish his territory, for example near exit/entrance doors

Place things where they have meaning in feline terms

Regularly introduce novel playthings and cater for mental and physical activity

Don't frighten your cats to discourage unwanted behaviour, they may claw more to leave reassuring scent

Don't ignore your cat, increased frustration may simply intensify the behaviour

Natural curiosity – domestic felines have flourished because they constantly investigate their environments looking for potential threats and opportunities. They know every inch of their territory and instantly spot changes. They examine every new item and, if it seems in any way challenging, they must make it smell familiar and reassuring. They do this by:

- Bunting or rubbing around an object to transfer their scent (see pages 30–31)
- Scratching as a visual and scent territory marker (see page 41)
- Occasionally spraying with urine (see pages 32–33)

None of this is naughty. There's a serious purpose behind such actions, but many puzzled owners watch in horror as their lovely new furniture, carpets or curtains come in for such undesirable attentions.

Suitable alternatives – some pets drive their owners crazy by shunning their toys and scratching facilities in favour of household furniture. This invariably relates to a lack of suitable items for scratching, playing and climbing on, or to their location in meaningless places, for example a scratching post that is placed in a little-used room.

Vocalization

Cats don't usually make much noise but some chat with their owners – a habit we invariably reward with fuss and attention. This can come back to haunt us! Cats are increasingly referred to behaviourists for vocal attention seeking.

WHAT IS HE TRYING TO SAY?
Your cat's meow could mean:

He's hungry

He wants some attention or a game

A female cat in 'heat' may call excessively

A new kitten may mew for attention, be frightened or missing his littermates

He may be disturbed by neighbourhood cats or generally agitated

Old age – elderly cats tend to be more vocal if they feel insecure

He may be unwell – seek your vet's advice if you think that there may be an underlying health problem

Excessive meowing – some cats are more talkative than others, and some breeds, such as the Siamese and raucously voiced Burmese, are known for their constant chatter. However, excessive meowing may indicate some underlying problems.

Wake-up call – very often cats are vocal in the early morning, and at bedtime, when cats are naturally up and about. Keep your cat out of your bedroom if you don't want him to invite you to join his morning patrol and provide some suitable nocturnal activities to keep him occupied, such as a puzzle or mechanical feeder that will dispense food at the time your cat wakes up, a box or bag with a ping pong ball inside to have fun with or a selection of toys.

Solutions – the difficulty is that by responding to your cat, you encourage his behaviour, as vocalization is often a means of attention seeking, whatever the cause. Bored

and under-stimulated cats will become a nuisance, especially if they are young, or mature but active and kept indoors without adequate entertainment. Ensure your cat is stimulated by:

- Providing a range of suitable, interesting exercise facilities and play items
- Using routine activities like feeding to add fun to your cat's life
- Giving plenty of the sort of attention your cat enjoys at appropriate times

ATTENTION SEEKING

Not all feline attention seeking is vocal. Clawing our prized possessions tends to get a reaction, too (see page 154). Tapping our sleeping faces, gently or with extended talons, is another early morning activity that tends to get a human response and not uncommonly cats try to join in with human activities whether their presence is welcome or not.

House soiling

Dirty habits are distressing. Inappropriate elimination and indoor marking can involve urine spraying and, rarely, middening (marking with faeces). Cats are never spiteful or obstinate, so look to their natural behaviour and your cat's needs to identify underlying causes and resolve problems.

What to do – once your veterinarian has ruled out or treated any contributing medical issues, distinguish between marking behaviour and inappropriate elimination. Beware – some cats do both at once and the evidence is sometimes confusing!

Marking – usually this takes the form of spraying a few drops of quite pungent urine several inches from the floor on vertical surfaces and objects. Cats may be seen in the typical spraying stance or intensely sniffing an introduced item, such as shopping bags, briefcases, buggies or new furniture, as a prelude to spraying.

Inappropriate elimination – cats often avoid going outside or using their litter trays if facilities are unsuitable, inappropriately located, not sufficiently clean or there is something frightening in the way (see pages 74–75). Clean affected locations well and look for physical or social causes to identify and rectify deficiencies. For example, cats will soil if:

- They do not like the tray design or litter type or it is not deep enough
- They cannot easily get to the tray or cat flap because it is too far away or someone has shut a door
- There is no suitable, secluded latrine area in the garden or the indoor tray is in a busy location
- The litter tray is not kept scrupulously clean
- A noisy toddler, boisterous, scary dog or intimidating feline housemate is blocking their way to the latrine

COMMON CAUSES OF MARKING

Stressed cats mark their core territory using urine, or faeces in rare cases, to make themselves feel more emotionally secure, but frustration can sometimes underlie such behaviour. Successful resolution depends upon identifying the causes of your cat's distress:

Timid or fearful cats with too few hiding places where they can de-stress

Cats that are pursued by people or other pets out of curiosity, for cuddles or games

Unrecognized tensions in multi-cat homes, especially when unrelated cats are treated as one social group. Providing plenty of separate dedicated facilities to reduce contact and competition can help (see page 82)

Changes in social composition. For example, a new partner, lodger, baby or pet, especially when their arrival coincides with cats being banned from areas of the home they used to frequent

Renovation and decorating that changes the physical and scent profile of the household

Aggression

Overt aggression – growling, hissing, spitting and swatting – is easily identified. Many cats only get that far if pushed to their limit, to extricate themselves from difficult situations or because past experience has taught them aggression gets what they want. Spotting lower-grade signs before it comes to this will help you resolve problems.

CALMING SOLUTIONS

Behaviour	Response
Avoids contact or becomes aroused or aggressive when you try to cuddle or pet him	Stop immediately. Talk to him all the time in a quiet, calm voice, respond with a very brief stroke to any overtures he makes to you or use play with a fishing-rod toy to show your affection. Cats should always be allowed to take the lead, even with familiar owners, or they may become self-defensive
Becomes aggressive with visitors, boisterous, intimidating children or dogs	Give your cat plenty of high and dark places to retreat to. These help cats to calm down, so clear the tops of your fridge freezer, kitchen units, cupboards and bookshelves, leave your airing cupboard and wardrobe doors open and provide hidey holes under beds
Tension between cats. If they avoid contact, sit and stare at one another or fight, particularly if they are not related, take active steps to reduce tension	Provide lots of beds, toys, and if they use litter trays, at least one each in different areas of your home, as well as splitting up their feeding stations so that they can avoid each other

Predatory behaviour is different from aggression. Cats need suitable outlets for their hunting skills, even when they aren't particularly skilled. Otherwise they may stalk people and other pets. (See also pages 36–37.)

What to do – never confront or punish an aggressive cat. Break eye contact because staring is a threat in feline terms. Slowly turn or back away. Attend to any injuries (people and pets) and use a thick blanket or sturdy gloves if you must handle your cat. Work out why he is aggressive and how to reduce his arousal. Get your veterinarian to check for illness or pain and seek professional help if your problem is significant and complicated.

Personality – consider your cat's personality, socialization history and experience. Timid or poorly socialized cats find busy homes stressful and intimidating, especially if people try to interact with them.

Competition for resources – and being forced together, particularly at mealtimes or in narrow spaces, commonly results in aggression in multi-cat homes. Be aware of signs of tension (see opposite) and don't overlook subtle body blocking by which one cat prevents another from reaching vital resources or the cat flap.

Feline fears

Although not the same, fear and anxiety are linked and may be the cause of disturbed behaviour. Many cats struggling with quite profound fears and anxieties simply retreat and hide so you may not recognize their distress.

GOLDEN RULES FOR FEARFUL CATS

Never make him 'face his fears' – this is unkind, will probably make things worse and could undermine your relationship

Don't cuddle or reassure him, just let your cat hide in one of your carefully provided cat refuges until he has settled

Remember your cat whenever something is anticipated at home – be particularly solicitous if he is timid, poorly socialized and elderly or has been frightened in the past

Take steps to protect him, for example, keep him safely inside when fireworks or storms are anticipated or put him in another room with a treat, some new toys and a cosy bed when unfamiliar visitors, especially children, are expected

Fear – is a negative emotional response to something or someone a cat regards as threatening. This may have little to do with them or their behaviour but rather relate to his timid personality, limited socialization experiences or subsequent unpleasant incidents. Understanding is essential.

Anxiety – is the unpleasant emotional state evoked by signs that something scary may happen. The classic example is hearing his cat basket being opened for the pet that has only ever been on stressful visits to the vet!

Phobias – sometimes cats find situations or stimuli so frightening they develop a phobia, a fearful state of such intensity they are no longer able to lead a normal life.

Stressors – circumstances vary but the range of feline stressors is wide. Some pets become agoraphobic (fear of open spaces) because they are unsociable with their own

kind and encounter many neighbouring cats; others avoid rooms where a smoke alarm went off ages ago; elderly cats become generally fearful; and a sensitive cat may run away when the doorbell goes because an over-friendly visitor insisted upon cuddling him in the past.

KEY TIP

If fear is negatively affecting your cat's life ask your veterinarian for a referral to a qualified feline behaviourist.

Signs of chronic stress

It is important to remember that the stress response is something that makes cats and people perform the necessary actions to avoid trouble. So long as they are then able to relax back into balance, good not harm is done. However, chronic, unrelieved stress can undermine emotional well-being and even physical health.

Compulsive disorders – repetitive or compulsive disorders are complicated feline medical conditions where stress and frustration are underlying factors. Affected cats repeat the same behaviour over and over as if they feel compelled to perform it. If you spot the signs early on, the chances of professionals offering effective help are greatly increased.

Psychogenic alopecia – normal grooming activity leaves a smooth, healthy coat but when very stressed some cats constantly wash and lick themselves, even pulling out tufts of fur, trying to calm themselves down. If clinical examination rules out parasites and other dermatological causes, a qualified behaviourist should be consulted.

Self-mutilation – unfortunately, some distressed cats also start to attack their tails. This is not the normal 'chasing the tail' game but rather an intense, frenzied action performed in a repetitive manner.

Feline orofacial pain syndrome (FOPS) – this recently identified syndrome is seen mainly in Burmese, although it can affect other breeds and non-pedigree cats. Exaggerated licking and chewing is seen and afflicted cats often claw intensely at their mouths. Urgent medical and behavioural attention is crucial.

Hyperaesthesia – also called 'twitchy cat disease' and 'rippling skin syndrome', this is characterized by dramatic skin twitching, bouts of frenzied dashing, as if cats are hallucinating, followed by freezing. It contrasts sharply with normal antics often referred to as a 'funny five minutes' when cats playfully rush about. Although the exact cause is unknown, it is understood that ongoing anxiety is a key factor.

Pica – Burmese and Siamese cats in particular can develop cravings for non-food items, usually fabrics containing wool or cotton. People sometimes find this funny but it should never be treated lightly. Affected cats need professional help to avoid a worsening of the compulsive behaviour and potentially life-threatening medical complications.

Signs of chronic stress

Problems outdoors

When pets go outside, the environment they encounter there is critically important. Making his outside territory as cat-friendly as possible is a great start, but you should also be aware of tensions, threats and stressors created by neighbourhood cats.

Outside freedoms – the great outdoors is the perfect place for your cat to be at his most natural, with plenty of opportunity to

hide, stalk, hunt, observe and relax. See pages 88–89 for providing a cat-friendly outside environment.

Boundaries – marking activities, such as scratching and spraying, should be permitted outside, preferably on untreated wooden posts. This will help your cat to advertise his territorial claims to other felines.

Nocturnal activities – cats like to patrol their territory at certain times of the day, and the evening is a natural hunting time. It is also a time that cats can come into conflict with each other. If your neighbourhood is home to several cats, then tensions may run high. To avoid potential clashes, and resulting injuries, you could try imposing a curfew on your cat, locking his cat flap and keeping him indoors late in the evening.

Unwanted visitors – you can also go some way towards limiting the incursions into his

territory of other cats by not encouraging them into your garden. In addition:

- Easy entrance points can be blocked
- Fences and walls extended using flimsy trellis to make access more difficult
- Non-harmful wire mesh barriers erected to ensure take-off areas like garages and shed roofs are off-limits

Such strategies can often help to reduce inter-cat battles as well as 'away days' that turn into much longer periods of absence.

PROBLEMS INDOORS

Behaviourists are often asked to assist with problems that relate to outdoor issues, although they occur only within the home. For example, some cats soil the house because their trips outside to relieve themselves are intimidating experiences. Creating a satisfactory outside space that caters well for cats can help avoid problems (see pages 88–89).

Treating problem behaviours

As we've seen, there is a considerable range of behaviours that can become problematic. Some are relatively straightforward to resolve, others more complex and challenging. Some behaviours need to be treated or diagnosed professionally.

Understanding – many of your cat's puzzling little habits can be simply explained by referring to natural feline behaviour. Looking at things from his point of view and making any necessary changes to his environment, your reactions and interactions, will deal with any problems effectively.

Professional help – it is important to take swift action if he starts to develop any serious behavioural issues. Check with your vet to ensure there is no underlying medical cause, and for referral to a well-qualified, experienced feline behaviourist, if necessary.

Treatment – depends upon not only establishing the underlying motivation but also identifying all possible contributing factors (see box). Once the problem is understood a behavioural modification programme can be started. This may include commercial pheromone products (see page 67), herbal remedies or psychoactive medication.

GETTING THE WHOLE PICTURE

Gather evidence, by keeping a diary of everything your cat does

Record your observations of his individual characteristics and reactions to everyday events, familiar and unfamiliar people and other pets, including cats

Note what goes on in the home and outside it – even if he is an indoor-only cat, he may be affected by seeing intruding cats through the window, hearing building noise or smelling challenging odours when you open a door or window

Draw a plan of your home. Detail your cat's facilities and locations where problems occur

Observe how you and others, two- or four-legged, react to your cat generally and specifically during problem times, plus any effects on his difficult behaviour

Senior care

Due to improved nutrition and better health care, cats are increasingly living to a ripe old age. Therefore, the chances of keeping your cat well into his teens have never been greater. You will need to monitor his health and make any changes to his lifestyle so that he continues to live in contentment.

Advancing years

To ensure he gets the best out of his 'retirement' you need to be aware of what happens during normal ageing and the changes you may need to make. Elderly felines are also prone to developing some specific diseases. It is important to look out for their clinical signs, because prompt treatment pays dividends.

Ageing gracefully – genetic inheritance, individual qualities, and the effects of nutrition, lifestyle and the environment mean cats age at different rates. However, there are three recognized stages of ageing –

mature, senior and geriatric – which will help guide you in their care (see table below).

'Winding down' – ageing occurs in all the body's cells, tissues and organs. The effects differ but most elderly cats suffer from:

- Deterioration in sight, hearing, their sense of smell and ability to taste
- Lower energy levels and changed activity patterns
- Orthopaedic conditions, such as muscular weakness and arthritis
- Immune compromise, which reduces resistance to infections and leads to slower recovery
- Greater sensitivity to cold, because they are less able to regulate body temperature
- Reduced ability to groom
- Weight change – some become increasingly frail, the waistlines of others dramatically expand

KEY QUESTION

Should I change my elderly cat's diet?

Elderly cats often develop medical conditions that are best managed by a change of diet and, as they age, cats generally need a lower protein diet. Always check first with your vet. Remember also that, as elderly pets deal less well than youngsters with change, they can find even minor variations in food and feeding routines upsetting. So you may need to be subtle, make changes gradually and, for instance, gently warm up meals to make them more aromatic, palatable and tempting. Take care if using a microwave as food may heat unevenly.

Mental changes – sadly, the brain does not escape the ravages of time and older cats invariably experience delays in processing information and responding. They can also have difficulties with perception and suffer from reduced concentration and memory loss. Learning isn't impossible for senior cats but they need time and patience and our expectations of them must always be realistic.

FELINE AGEING STAGES

Mature	7–10 years
Senior	11–14 years
Geriatric	15+ years

Problem signs

It is important to avoid attributing increased lethargy, lack of agility and changes in weight or body shape, irritability and confusion to 'just growing old'. These can all be subtle indicators that a cat is suffering from a medical condition and needs help.

Normal changes – elderly cats commonly sleep more during the day but they may become wakeful and noisy at night. They can also be slower 'coming to' when they wake, taking a few minutes to orientate themselves and recognize familiar people and situations.

Regular check-ups – as your cat enters his senior years, there is considerable value to be gained from visiting your vet at least twice a year, although more frequent checks are often necessary as time goes on. They will record baseline data, such as weight, liver, heart and kidney function, for future reference and to monitor changes. Many conditions are more effectively treated, even when cure is not possible, if they are identified early.

Increased anxiety – an especially important issue for any owner to bear in mind is that older cats are more prone to anxiety- and fear-related conditions. This makes their behaviour somewhat inflexible and they can be resistant to change, so sensitive handling, particularly at times of stress or upheaval, is essential (see pages 146–147).

DAILY CHECKS

Any change, for example apathy, hiding or reluctance to move, in an older cat's established patterns of behaviour may be suspicious. Contact your vet if:

Thirst increases or decreases

Appetite is abnormal

Weight changes up or down

Body shape is altered

Difficulty is encountered with eliminating – elderly cats often suffer from constipation but diarrhoea needs attention, too

---

Age-related diseases

Cats are subject to a range of medical conditions and diseases. Some potentially life-threatening illnesses affect older cats in particular and knowing their associated signs will help you to identify them quickly and immediately seek veterinary help.

COMMON FELINE DISEASES IN OLDER CATS

Disease	Symptoms	Treatment
Dental disease	May cause anorexia (loss of appetite) or discomfort when cats eat. Signs include wincing, pawing the mouth or chewing on one side, halitosis (smelly breath), dribbling and bleeding	Scaling and extractions under general anaesthetic with medication for gum disease
Kidney disease	Chronic renal failure is common. Increased thirst, changes in appetite, especially anorexia, and weight loss are usually seen; often accompanied by lethargy and confusion as the condition progresses	Supportive medication and restricted protein and phosphate diets to help manage the condition for which there is no cure
Hyperthyroidism	Increased thyroid hormone in the blood causes changes in 'personality' with even sociable cats becoming irritable and aggressive, disorientated and hyperactive. They quickly lose weight and often have diarrhoea	Medication, surgery or radioactive iodine (in specialist centres)
Diabetes mellitus	Increased thirst, weight loss, depression in the early stages; progressive blindness and organ failure if untreated. Diabetic animals are also prone to developing infections	Specialized diets, insulin injections and sometimes other medication
High blood pressure and heart disease	May go unnoticed until serious symptoms are apparent, such as intraocular haemorrhage (bleeding into the eyeball) or thrombosis. More vets routinely check cats' blood pressure and heart function during senior pet health check-ups	A number of effective drugs are now used to control feline hypertension and support cats with cardiac problems
Cancer	Cats are predisposed to a number of neoplastic conditions. Any tissue or organ can be affected but cats with white ears and noses may develop skin malignancies, while oral and abdominal tumours are not uncommon. Signs depend upon the nature of each cat's disease but often include appetite and thirst changes, weight loss and diarrhoea	When this is possible it involves surgery, chemotherapy or a combination of the two

Feline dementia

Only recently recognized, Cognitive dysfunction syndrome (CDS) is the feline equivalent of Alzheimer's disease in people. Sadly, no cure exists but supportive treatments can sometimes help to improve the quality of life of affected cats.

CLINICAL SIGNS OF CDS

The acronym DISHA is commonly used:

Disorientation – cats become easily confused even in familiar places

Inhibition – some pets seem to have personality changes and no longer recognize other household members, people or pets, to whom they were previously well bonded; others become more 'clingy'

Sleep/wake cycle changes – excessive vocalization and night-time pacing are seen in some individuals

House soiling – but all the other possible causes must be ruled out or dealt with before a diagnosis of CDS is made

Activity reduction and anorexia – can feature in the changes some pets exhibit, for example they may become very lethargic and stop grooming

Early signs – this progressive condition results in deterioration of brain function with the rate of decline varying between individuals. Unfortunately, in the initial stages clinical signs can be subtle and difficult to distinguish from normal ageing change (see box). This matters because the earlier supportive therapies are introduced the better the results are likely to be. However, knowing what may happen can often help owners to pick up behavioural indicators that an elderly cat may be developing CDS. As no specific test is available, diagnosis is made by excluding other conditions.

Helpful measures – making life as simple as possible for affected cats can help avoid unnecessary confusion and lessen the effects of the various problems that arise as a result of deteriorating mental powers. Such cats often benefit from having access to only a small area of the home, for instance. It is

also important to ensure that furniture and routines stay the same in order to help them cope better. If change is essential, for example moving house, minimising changes that affect the cat or making alterations in a staged manner is ideal. In addition, veterinarians can prescribe diets that are specially enriched with anti-oxidants, or supplements if a cat resists such dietary change. Medications that have a neuroprotective function and enhance blood flow to the brain and those that increase the activity of some neurotransmitters are also used to support patients suffering from CDS.

Good quality of life

Recognition and treatment of pain and quality of life are very much bound together. Fortunately, advances in veterinary medicine mean that there are now more drugs available to help reduce the effects of painful conditions in cats.

Be alert – sensitive observation is even more important as cats age and knowing your cat's well-established behaviour patterns will be an advantage when spotting signs of illness or physical difficulty. Change is significant – watch for subtle signs, as felines tend to

INDICATORS OF PAIN	
General demeanour and temperament	Agitation, restlessness, withdrawal, avoidance, attempts to escape, hiding, aggression
Posture	Hunched, slumped, drooping head, lying flat, tucked up abdomen
Locomotion	Reluctance to move, lameness, unusual gait
Vocalization	Hissing, spitting, crying, moaning, purring, yowling
Loss of self-maintenance	Changes in eating, drinking, grooming, eliminating habits
Physical	Increased respiratory rate, dilated pupils
Activity	Depressed, increased or changed. For example, disrupted sleep patterns
Social interaction	Altered, especially if sociable cats become irritable and 'anti-social'

SOILING

House soiling may not be a sinister development but may simply reflect a cat's inability due to pain and immobility to get to his latrine area in time. Or he may be reluctant to go outside to relieve himself because of nasty weather or an intimidating feline neighbour.

suffer in silence. For example, many cats become less agile as a result of arthritis and its associated discomfort. If you notice your cat avoiding his previously favoured elevated sleeping places, get him checked out and try some supportive aids, such as steps, to get up there (see pages 184–185). Make an extra effort if yours is a very independent cat and you don't see much of him. He may drink more outside, be constipated or lethargic without you realizing.

Assessing pain – it is very difficult to tell whether a cat is in pain. During a clinical examination a vet will try to identify physical abnormalities and assess a cat's reaction to palpation – wincing or hissing when a particular area is touched is significant. There are other indicators that you can look out for – see the box opposite.

Senior sensitivities

When he reaches his senior years, in order to continue his good quality care and allow him to behave as his species and individual characteristics dictate, you need to be aware of any lifestyle changes that may cause distress, adjusting conditions as required.

KEY QUESTION

How can I help my elderly cat adjust to a new kitten in the household?

It is important to handle things carefully, making sure that your cat is not bothered or stressed by the newcomer. Keep his favourite spots and his routines the same as before. Create an interesting and entertaining environment for the youngster – this is essential, as bored kittens often pursue and distress older cats. After all, what is more entertaining than a rapidly retreating tail? Make sure that you provide many more beds, hiding places, toys plus extra litter trays in different locations so each cat can have privacy. Feed your cats in completely different areas, as mealtimes are often a source of competition and stress, and your cats' nutritional needs will be completely different. Never leave them alone together unless and until you are certain they are living harmoniously.

Sensitive natures – increased anxiety, reduced adaptability and resistance to change mean that even minor variations in their physical or social worlds can have a profound emotional impact on our elderly cats.

TIPS FOR COPING

Introduce change gradually

Give your cat access to rooms and possessions that retain the scent of their special person, if they are absent for some time

Use commercial pheromone products to alleviate stressful situations (see page 67)

Use your cat's own scent to introduce new items of furniture or help with trips away from home

Change – the loss of human companions and other pets, additions to the household or alterations within it can cause confusion and distress, especially when familiar and reassuring scents are lost. Moving home can be especially disorientating, but awareness of the issues and good planning can help you to reduce the negative impact of these common stressors upon an older pet (see pages 146–147).

Lifestyle changes

An elderly pet requires a few adjustments to his everyday activities. By being aware of practicalities in feeding, sleeping and grooming, and paying attention to maintaining routines and giving him opportunities for both physical and mental activity, you will make his senior years easy-going and enjoyable for you both.

Practical solutions – try to encourage reasonable physical effort without overtaxing a less active or increasingly infirm cat. Place his feeding and toileting facilities near to his favourite sleeping and resting areas, particularly if your home is spacious. In fact, it is sometimes necessary to provide duplicate sets of beds, litter trays, feeding stations and water points in various areas so that they are readily to hand when an older cat needs them.

> **KEY TIP**
>
> Dehydration in elderly cats is becoming a cause for concern among vets (see pages 136–137). To avoid this problem be aware of where he likes to drink, the type of container he prefers, and make sure he can easily reach his water source.

Modifications – equipment may need to be modified and aids provided in the form of ramps or steps for elderly cats to get up to safe refuge areas. As your cat ages consider:

- He may need low-sided, larger litter trays for easy access and with enough room for a less agile, arthritic cat to turn around comfortably
- A variety of cosy, draught-proof beds with extra padding especially if he's becoming frail, as pressure on his joints may be increasingly painful
- A number of easily accessed refuges in busy areas of the home where proximity to human activity or other pets may be especially stressful
- Use of different textures, scents, for example lavender oil, and sounds to differentiate various areas of the house for cats that are losing their sight or mental acuity

- A pet fountain which helps partially sighted cats locate their water
- Easy puzzle feeding activities and interactive play to encourage gentle exercise and mental activity
- Suitable toys, such as light objects with bells, for the same purpose
- Avoid moving furniture in areas where your cat spends his time. Even minor changes may upset him
- Prevent inadvertent pressure from visitors who want to interact with him. Elderly cats are less likely to remember people they see only occasionally than when they were younger, so their attentions may cause unintentional distress

Nearing the end

Sadly, the time inevitably comes when a pet's quality of life is questionable and you need to consider the option of euthanasia. Coping with this and the loss of your pet is a difficult time but being prepared can help you make the right decisions.

Action not regret – no matter how hard it is, tough decisions must be made to avoid unnecessary suffering. Letting go of a treasured cat is the loving thing to do.

Planning ahead – as your cat ages, discuss what is involved in euthanasia with your vet on a routine visit when you can think clearly.

Euthanasia – sometimes referred to as 'being put to sleep', this involves an injection, which causes his heart to stop, usually given via a leg vein. The process takes just a few seconds: during and after the injection the cat becomes drowsy, lapses into unconsciousness and dies peacefully.

At the end – some clinics offer house visits but many owners prefer to take their pets to the clinic. Unless something dramatic happens, arrange an appointment at a quiet time to avoid delay. Stay with your cat if you want to and can control your emotions

enough not to upset him, but doing what is best for your pet is of paramount importance.

Afterwards – examine the options of routine or individual cremation, burial in a pet cemetery or at home.

BEREAVEMENT

For everyone in a household, pets and people, the loss of a much-loved cat can be upsetting. However, think carefully before rushing out to acquire another cat. It is wise to mourn the cat that has gone and adjust to changed circumstances, no matter how long it takes. Then, at a later date, get another cat, if it really is a desirable and sensible thing to do. If the loss seems unbearable, there are organizations that offer bereavement support and others that can advise on how to help grieving pets come to terms with the deficit in their social group.

Index

Acknowledgements

I would like to thank everyone at Hamlyn who helped to produce this book, especially Trevor Davies, Charlie Hallam, Katie Hardwicke and Charlotte Macey, my editor, with whom it has been a pleasure to work. Naturally thanks must also go to all the cats it has been my privilege and pleasure to know – and learn from – over the years.

Executive Editor Trevor Davies
Senior Editor Charlotte Macey
Executive Art Editor Penny Stock
Designer Pete Gerrish
Senior Production Controller Carolin Stransky
Picture Researcher Emma O'Neill

Alamy Adrian Sherratt 176; F1online digitale Bildagentur GmbH 91; Geoffrey Robinson 1; Isobel Flynn 66; 141; Juniors Bildarchiv 18; 33; 97; 137; 143; 154; 158; Mark Bond 170; Ovia Images 61; Picture Partners 2; Rob Walls 40; www.ourwildlifephotography.co.uk 22

Ardea John Daniels 75

Corbis David Woods 7; Jane Burton/Dorling Kindersley 165; K & H Benser 27; Philip Harvey 86; Photo 24/Brand X 71

Dorling Kindersley Dave King 50

FLPA Imagebroker 185

Fotolia EcoView 10; Hiro 65; Katerina Cherkashina 49; Robert 88; Tony Campbell 4

Getty Images DAJ 157; Dorling Kindersley 77, 79; Duncan Smith 153; GK Hart/Vikki Hart 43, 138; Image Source 148; John Kelly 85; Jose Luis Pelaez Inc 183; Julia Kuskin 145; Kevin Steele 53; Lisa Valder 169; Michael Engman 15; MIXA 44; Neo Vision 81; Petography 150; Robert Stahl 39; Sami

Sarkis 147; Steve Gorton & Tim Ridley 92; Vincenzo Lombardo 112

Masterfile 54; Jerzyworks 83

Nature Picture Library De Meester/Arco 101; Jane Burton 20; Wegner/Arco 74

Octopus Publishing Group 28, 115, 123, 124, 126, 128-9, 130; Jane Burton 47, 62, 73, 95, 99, 107, 110, 122, 133; Keith Colin 120, 121; Ray Moller 51; Steve Gorton 24, 30, 166, 187

Photolibrary Aflo Foto Agency 59; Age Fotostock/Pedro Coll 89; Juniors Bildarchiv 103, 134, 163; Mixa Co Ltd 56; Morales Morales 35; Vstock 175; WestEnd61 8; 37

Photoshot Yves Lanceau 48

RSPCA Angela Hampton 117

Tips Images Arco Digital Images 84, 161, 179, 181; F1 Online 69; ImageSource 167

Warren Photographic 104, 119, 173